Fractal Speech Processing

Although widely employed in image processing, the use of fractal techniques and the fractal dimension for speech characterization and recognition is a relatively new concept, which is now receiving serious attention. This book represents the fruits of research carried out to develop novel fractal-based techniques for speech and audio signal processing. Much of this work is finding its way into practical commercial applications.

The book starts with an introduction to speech processing and fractal geometry, setting the scene for the heart of the book where fractal techniques are described in detail with numerous applications and examples, and concludes with a chapter summing up the potential and advantages of these new techniques over conventional processing methods. It will provide a valuable resource for researchers, academics and practising engineers working in the field of audio signal processing and communications.

Professor Marwan Al-Akaidi is Head of the School of Engineering and Technology at De Montfort University, UK. He is a Senior Member of the Institute of Electrical and Electronic Engineers and Fellow of the Institute of Electrical Engineering. He is Chair of the IEEE UKRI Signal Processing Society and has presided over many national and international conferences in the field.

Fractal Speech Processing

Marwan Al-Akaidi

PUBLISHED BY THE PRESS SYNDICATE OF THE UNIVERSITY OF CAMBRIDGE
The Pitt Building, Trumpington Street, Cambridge, United Kingdom

CAMBRIDGE UNIVERSITY PRESS
The Edinburgh Building, Cambridge CB2 2RU, UK
40 West 20th Street, New York, NY 10011–4211, USA
477 Williamstown Road, Port Melbourne, VIC 3207, Australia
Ruiz de Alarcón 13, 28014 Madrid, Spain
Dock House, The Waterfront, Cape Town 8001, South Africa

http://www.cambridge.org

First published 2004

Printed in the United Kingdom at the University Press, Cambridge

Typefaces Times 10.5/14 pt. and Helvetica *System* LaTeX 2_ε [TB]

A catalogue record for this book is available from the British Library

Library of Congress Cataloguing in Publication data

Al-Akaidi, Marwan, 1959–
Fractal speech processing / Marwan Al-Akaidi.
p. cm.
Includes bibliographical references and index.
ISBN 0-521-81458-8
1. Speech processing systems. 2. Fractals–Data processing. I. Title.
TK7882.S65A43 2004
006.4′54 – dc22 2003055750

ISBN 0 521 81458 8 hardback

Contents

Acronyms and abbreviations

AbS	analysis by synthesis
ADM	adaptive delta modulation
ADPCM	adaptive differential pulse code modulation
AUSSAT	Australian Satellite
BA	binary arithmetic
BCM	box-counting method
BPP	bounded-away error probabilistic polynomial-time computations
CA	cellular automata
CCITT	International Consultative Committee for Telephone and Telegraph
CDF	cumulative distribution function
CELP	code-excited linear prediction
CM	correlation method
CPU	central processing unit
DAC	digital-to-analogue converter
DCC	discrete chaotic cryptology
DCT	discrete cosine transform
DFT	discrete Fourier transform
DM	delta modulation
DP	dynamic programming
DPCM	differential pulse code modulation
DSP	digital signal processing
DTMF	dual-tone multifrequency
DTW	dynamic time warping
EEG	electroencephalogram
FDS	fractal-dimension segmentation
FFT	fast Fourier transform
FIR	finite impulse response
FPA	floating-point arithmetic
GSM	Global System for Mobile Communication
HMM	hidden Markov model
IFFT	inverse fast Fourier transform

IMBE	improved multiband excitation
INMARSATM	International Maritime Satellite
IFS	iterated-function system
KLT	Karhunen–Loeve transform
KS	Kolmogorov–Sinai (entropy)
LCG	linear congruential generator
LFSR	linear feedback shift register
LP	linear approximation probability
LPC	linear predictive coding
MAP	maximum *a posteriori*
MBE	multiband excitation
ML	maximum likelihood
MMEE	minimum mean square error estimator
MPC	multipulse coding
NFSR	non-linear feedback shift register
PC	polynomial-time computations
P-box	permutation box
PCM	pulse code modulation
PCNG	pseudo-chaotic number generator
PDF	probability density function
PNP	non-deterministic polynomial-time computations
PRNG	pseudo-random-number generator
PSDF	power spectral density function
PSM	power spectrum method
PSTN	Public Switched Telephone Network
RELP	residual excited linear prediction
RPE	regular-pulse-excitation coder
RSF	random scaling fractal
S-box	substitution box
SFD	stochastic fractal differential
SME	spectral magnitude estimation
SNR	signal-to-noise ratio
STC	sinusoidal transform coder
STFT	short-time Fourier transform
VQ	vector quantization
WDM	walking-divider method
XOR	'exclusive or' gate

1 Introduction to speech processing

1.1 Introduction

Speech is and will remain perhaps the most desirable medium of communication between humans. There are several ways of characterizing the communications potential of speech. One highly quantitative approach is in terms of information theory. According to information theory, speech can be represented in terms of its message content, or information. An alternative way of characterizing speech is in terms of the signal carrying the message information, that is the acoustic waveform [1].

The widespread application of speech processing technology required that touch-tone telephones be readily available. The first touch-tone (dual-tone multifrequency, DTMF) telephone was demonstrated at Bell Laboratories in 1958, and deployment in the business and consumer world started in the early 1960s. Since DTMF service was introduced to the commercial and consumer world less than 40 years ago, it can be seen that voice processing has a relatively short history.

Research in speech processing by computer has traditionally been focused on a number of somewhat separable, but overlapping, problem areas. One of these is isolated word recognition, where the signal to be recognized consists of a single word or phrase, delimited by silence, to be identified as a unit without characterization of its internal structure. For this kind of problem, certain traditional pattern recognition techniques can be applied directly. Another problem area is speech coding and synthesis, where the objective is to transform the speech signal into a representation that can be stored or transmitted with a smaller number of bits of information and later restored to an intelligible speech signal.

Speech coding or speech compression is the field concerned with obtaining compact digital representations of speech signals for the purpose of efficient transmission or storage. Speech coding uses sampling and amplitude quantization. The sampling is done at a rate equal to or greater than twice the bandwidth of analogue speech (according to Nyquist theory). The aim in speech coding is to represent speech by a minimum number of bits while maintaining its perceptual quality. The quantization or binary representation can be direct or parametric. Direct quantization implies binary representation of the speech samples themselves. Parametric quantization uses binary representation of the speech model and/or spectral parameters. The simplest non-parametric coding technique is pulse code modulation (PCM), which is simply a quantizer of sampled amplitudes. Speech coded at 64 kbps using logarithmic PCM is considered as non-compressed and is often used as a reference comparison [1–3]. Quantization methods that exploit signal correlation, such as differential PCM (DPCM), delta modulation (DM) [4] and adaptive DPCM (ADPCM) were proposed later, and speech coding with PCM at 64 kbps or ADPCM at 32 kbps became the CCITT standard [2, 3].

Speech-specific coders or voice coders (vocoders) rely on a linear speech-source–system-production model [1]. This model consists of a linear slowly time-varying system (for the vocal tract and the glottal model) excited by periodic impulse train excitation for voiced speech and random excitation for unvoiced speech. In this model, the vocal tract filter is all-pole and its parameters are obtained by linear prediction analysis, a process in which the present speech sample is predicted by a combination of previous linear speech prediction techniques [5]. Atal and Hanauer [6] reported an analysis-by-synthesis system based on linear prediction.

Homomorphic analysis is another method for source-system analysis. It is used for separating signals that have been combined by convolution. One of the advantages of homomorphic speech analysis is the availability of pitch information from the cepstrum [7]. Schafer and Rabiner [8] designed and simulated an analysis-by-synthesis system based on the short-time Fourier transform (STFT), and Portnoff [9, 10] provided a theoretical basis for the time–frequency analysis of speech using the STFT. Research in linear prediction, transform coding and subband coding continued; Tribolet and Crochier [11] proposed a unified analysis of transform and subband coders. A federal standard, FS-1015, based on the LPC-10, was developed in the early 1980s [12]. In the 1980s and 1990s research was focused on developing robust low-rate speech coders capable of producing high-quality speech for communications applications. Much of this work was driven by the need for narrowband and secure transmission in cellular and military communications.

In the 1980s many types of speech coder were developed: coders using sinusoidal analysis-by-synthesis of speech, proposed by McAulay and Quatieri [13, 14]; multiband excitation vocoders, proposed by Griffin and Lim [15]; coders using multipulse and vector excitation schemes for linear predictive coding (LPC), proposed by Atal and Remde [16]; vector quantization (VQ) coders, promoted by Gerson and Jasiuk [17]. Vector quantization proved to be very useful in encoding LPC parameters. In particular, Schroeder and Atal [18] proposed a linear prediction algorithm with stochastic vector excitation called code-excited linear prediction (CELP). The stochastic excitation in CELP is determined using a perceptually weighted closed-loop (analysis-by-synthesis) optimization. The CELP coder is called a hybrid coder because it combines the features of traditional vocoders with waveform features. The real-time implementation of hybrid coders became feasible with the development of highly structured codebooks.

Progress in speech coding has enabled the recent adoption of low-rate algorithms for mobile telephony. An 8 kbps hybrid coder has already been selected for the North American digital cellular standard [17], and a similar algorithm has been selected for the 6.7 kbps Japanese digital cellular standard. In Europe, a standard that uses a 13 kbps regular pulse-excitation algorithm [19] was completed and partially deployed by the Global System for Mobile Communication (GSM). Parallel standardization efforts for secure military communications [20] have resulted in the adoption of a 4.8 kbps hybrid algorithm for Federal Standard 1016 [21]. In addition, a 6.4 kbps improved multiband excitation coder has been adopted for the International Maritime Satellite (INMARSATM) system [22, 23] and the Australian Satellite (AUSSAT) system.

1.2 The human vocal system

Speech can be defined as waves of air pressure created by airflow pressed out of the lungs and going out through the mouth and nasal cavities. The air passes through the vocal folds (chords) via the path from the lungs through the vocal tract, vibrating them at different frequencies (Figures 1.1, 1.2).

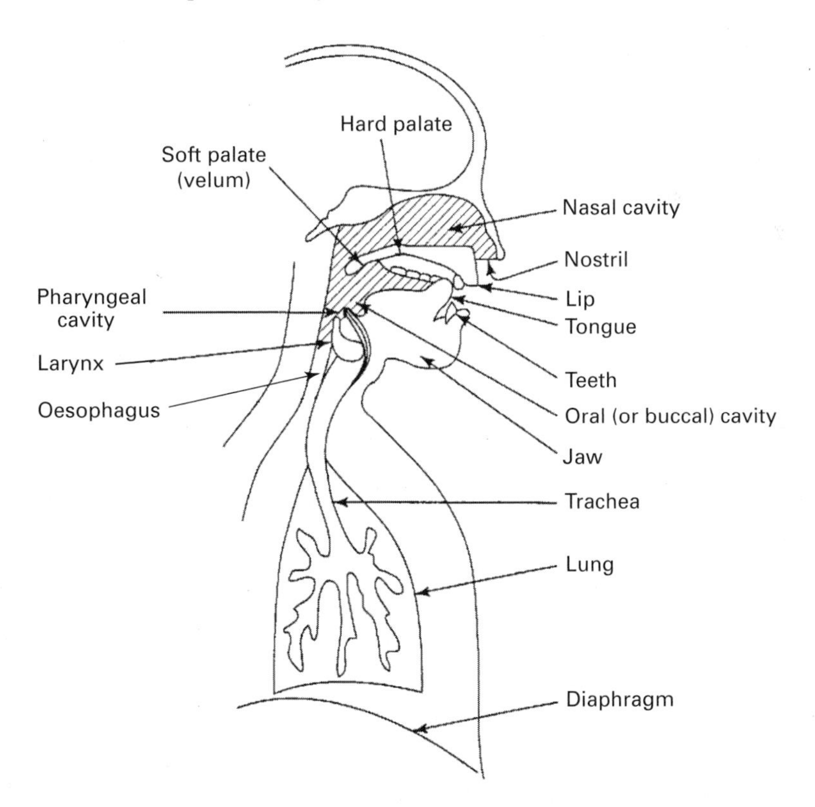

Figure 1.1 The human articulatory apparatus.

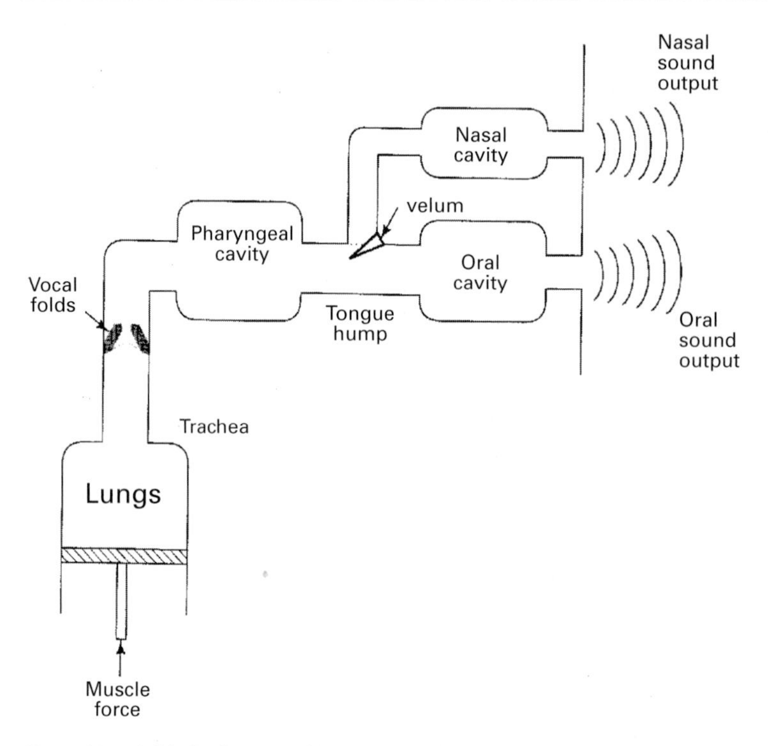

Figure 1.2 A block diagram of human speech production [24].

The vocal folds are thin muscles looking like lips, located at the larynx. At their front end they are permanently joined, and at the other end they can be open or closed. When the vocal folds are closed, they are expanded against each other, forming an air block. Air under pressure from the lungs can open that air block, pushing the vocal folds aside. The air passes through the crack thus formed and the pressure declines, allowing the vocal folds to close again. This process repeats, vibrating the vocal folds to give a voiced sound. The vocal folds in males are usually longer than in females, causing a lower pitch and a deeper voice.

When the vocal folds are open, they allow air to reach the mouth cavity easily. Unvoiced sound is formed by a constriction in the vocal tract, causing air turbulence and then random noise.

The resonance frequencies of the vocal tract tube are called the formant frequencies or simply the formants [1]. The vocal tract is a tube of non-uniform cross section and is about 17 cm long in adult males, usually open at one end and nearly closed at the other. It branches nearly at its midpoint to give the nasal cavity, a tube about 13 cm long, with a valve at the branch point (the velum), as shown in Figure 1.3. If the velum is closed, this excludes the nasal cavity from consideration and greatly simplifies analysis. If the vocal tract were of uniform cross section, many natural frequencies would occur at

$$f_n = (2n - 1)\frac{c}{4L} \tag{1.1}$$

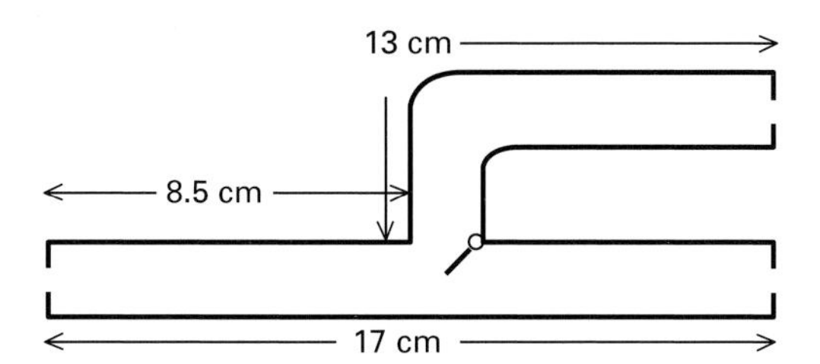

Figure 1.3 Simplified diagram of the vocal tract. The straight section reaches from the glottis at the left-hand end to the lips at the right-hand end, and the curved section reaches from the velum to the nostrils [25].

where $n = 1, 2, 3, \ldots$ In air $c = 350$ m/s, so, for a tube of length $L = 17$ cm, these frequencies occur at odd multiples of almost 500 Hz. In fact, the cross section is not uniform; as a result the resonances are not equally spaced, but the average density of vocal tract resonances is still approximately one per kHz of bandwidth, as the above relation suggests [25].

1.3 Acoustic phonetics

Speech contains vowels and consonants. All the vowels are produced when the vocal folds are closed (e.g. i(IY), a(A), o(O)), therefore they are voiced. The consonants, however, can be voiced, unvoiced or semi-voiced. Voiced consonants are formed by closing the vocal folds (e.g. m(M), n(N)). For unvoiced consonants (e.g. s(S), f(F)) there is no vocal-folds vibration, because they are open and the excitation in this case is the turbulent flow of air passing through a constriction in the vocal tract produced by the tongue. The semi-voiced consonants (e.g. v(V), z(Z)) are a combination of these two, for example v(V) is said like f(F) but with the vocal folds closed.

1.4 The fundamental speech model

The speech mechanism can be modelled as a time-varying filter (the vocal tract) excited by an oscillator (the vocal folds), with different outputs; see Figure 1.4. When voiced sound is produced, the filter is excited by an impulse sequence. When unvoiced sound is produced, the filter is excited by random white noise, which includes hardly any periodicity [3, 26, 27].

Signal representation

A speech signal can be represented in a digital form. This signal is represented as a file in the computer. In the present text we will deal with sound files in WAV format.

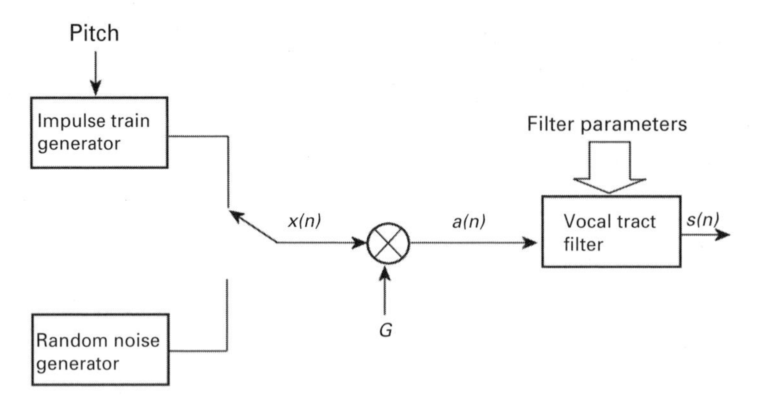

Figure 1.4 Block diagram of the source-filter model of speech: G is the gain, $x(n)$ the input signal, $a(n)$ the amplified signal and $s(n)$ the output signal; n is a digitized time variable.

The MATLAB mathematical system can read, write and play these files. This file has the following main features:

- file name
- sampling frequency (in Hz)
- resolution or number of quantization levels (in bits)

The sound is represented as a digitized signal, with an amplitude that changes in value between -1 and $+1$. This range is splintered into 256 levels if the resolution is 8 bit and into 65 536 levels if the resolution is 16 bit. The sampling frequency can have various values and we will fix our view about it in special cases.

1.4.1 The DC component

Let us consider a particular speech signal, the English word 'majesty'. Usually, speech recorded in real conditions from a microphone has a constant component, called the DC component in electronics. As explained below, it is essential that this component is removed.

There are two methods of removing the DC component. In method (i) the forward fast Fourier transform (FFT) is generated, then the first frequency component is cleared and finally the inverse FFT (IFFT) is taken.

Example of method (i) Let x be the original speech signal, which contains a DC component. Calculate y, which is the signal without the DC component. This is shown in Figure 1.5. The code follows:

```
xf=fft(x);        % forward FFT
xf(1)=0;          % clear the first component
xc=ifft(xf);      % inverse ???
y=real(xc);       % take only real part
```

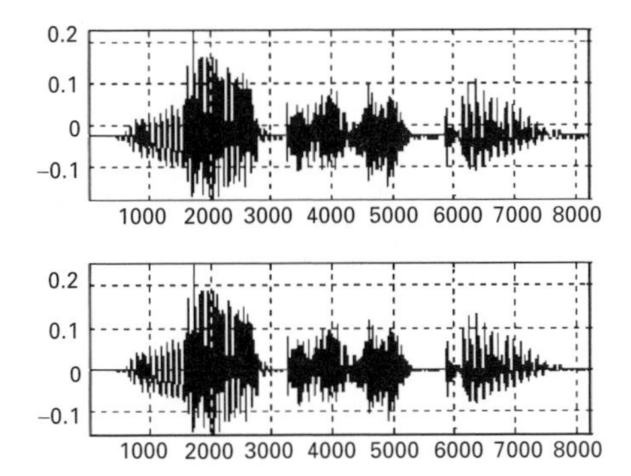

Figure 1.5 Original signal for the word 'majesty' (upper figure) and signal without DC component (lower figure).

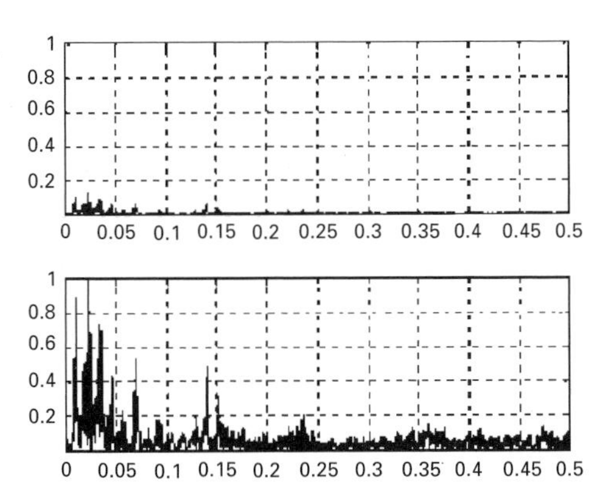

Figure 1.6 Spectrum of the original signal (upper figure) and after the DC has been cleared (lower figure).

The removal of the DC component is very useful for spectrum visualization. The fact is that the corresponding first frequency element in the Fourier transform usually has a large value. In this case, the corresponding spectral value will be a high peak. The useful data, however, will have low spectral values and is weak in comparison with the spectral peak of the DC. Clearing the DC component gives us magnification of the useful spectral peaks. This situation is illustrated in Figure 1.6.

Method (ii) involves subtraction of the DC from the signal. The DC component is offset relative to zero. It is calculated as the mean of the signal. For example, if the signal is represented by the number −0.0256, it is situated below the zero axis. In other words, the signal has a negative offset. By subtracting the offset from the signal we return the signal to zero.

Example of method (ii) Let x be the original speech signal, which contains DC, and let L be the length of the samples. Let us calculate y, which is the signal without the DC component:

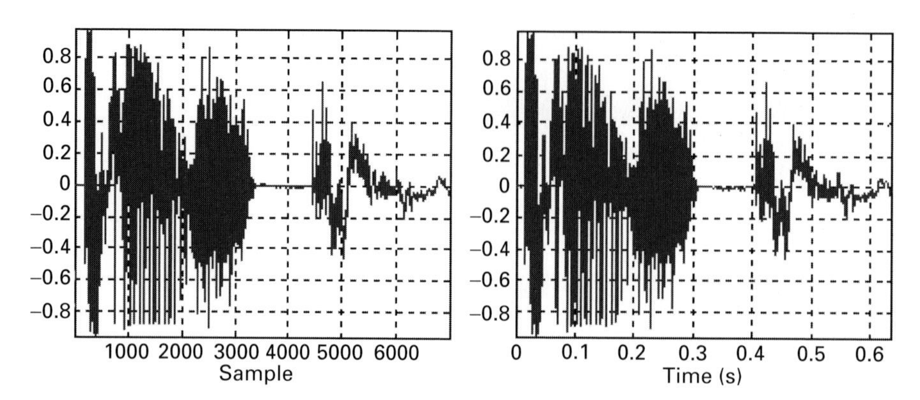

Figure 1.7 Waveform of the word 'test'.

```
ofs=mean(x);   % calculate mean
y=x-ofs;       % correct the offset
```

As discussed above, removing the DC component makes a spectrum more readable. This method is better than the previous one, because it is not necessary to calculate the forward and inverse FFTs, which needs a considerable amount of calculation. In what follows, we will deal only with the signal without DC component.

1.4.2 Waveform signal representation

A speech signal can be represented as a waveform, as shown in Figure 1.7. This file is characterized by the following features:

- the sampling frequency is 11 025 Hz;
- the resolution is 8 bit (256 levels of quantization);
- the length L of the file is 7000 samples.

The waveform shown on the left-hand side has as its horizontal axis the sample number: (it is therefore called the sampling axis). The same waveform is shown on the right-hand side with the time variable along the horizontal axis. We can calculate the period of a single sample if the sampling frequency is known:

$$t_s = \frac{1}{f_s} = \frac{1}{11\,025\ \text{Hz}} = 90.7\ \mu\text{s}$$

It is possible to label the horizontal axis in seconds if the sampling period is known. The total playing time of the file is calculated as the product of the file length (in samples) and the sampling period (in seconds):

$$T_s = Lt_s = 7000 \times 90 \times 7\ \mu\text{s} = 0.635\ \text{s}$$

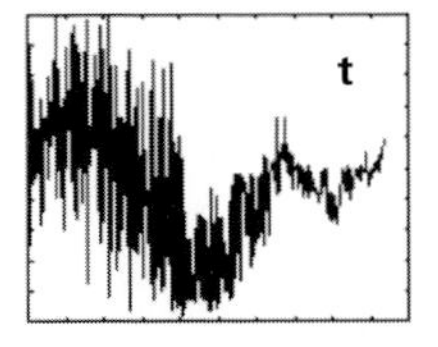

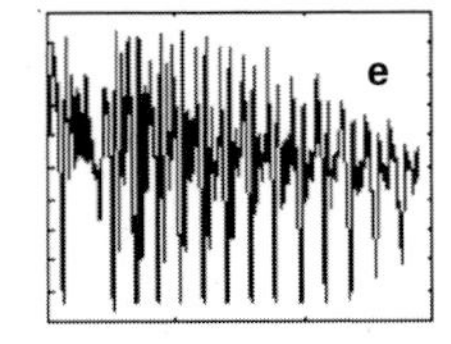

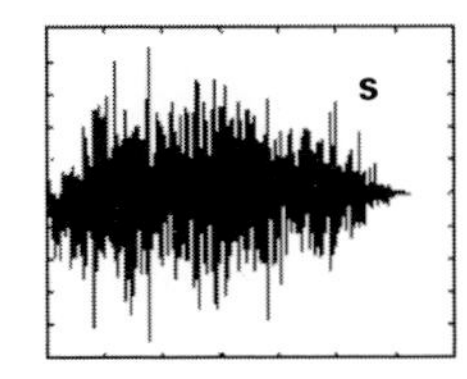

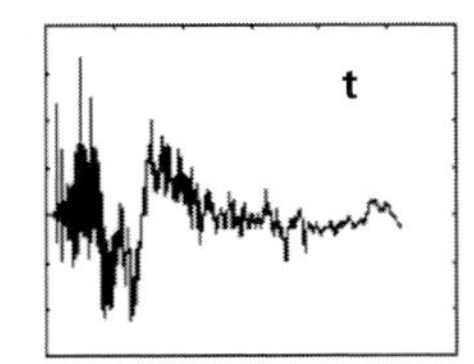

Figure 1.8 The four sounds of the word 'test': 't', 'e', 's', 't'.

What information can the waveform give us? Let us look carefully at Figure 1.7. In this example, the word 'test' was pronounced. This word consists of four sounds: 't', 'e', 's', 't'. It is possible to recognize a small pause before the last letter 't', because the word was spoken slowly and distinctly. Thus, the word was pronounced as 'tes–t'. If we look carefully at the waveform, we can visually recognize some sounds. There are four independent sounds in this waveform. Therefore, the original waveform can be split as shown in Figure 1.8.

The following can be concluded.

1. The consonants usually have a 'noisy' structure while the vowel 'e' has a periodic nature.
2. The first and the last waveform, each representing the letter 't', are similar and both have an explosive character. However, the first waveform corresponds to open vocal folds and the last to closed vocal folds.
3. The letter 's' appears like white noise.

Thus, by inspecting the waveform, it is possible to draw a conclusion about the type of sound. We can identify consonants or vowels (noisy or periodic) and explosive or monotonous (short or long) sounds. But how accurate is the segmentation achieved by using waveforms only? What are the available methods which help us to make true segmentation and recognition?

Waveforms are in the time domain (amplitude–time), but it is important to look at a waveform also in the frequency domain (amplitude–frequency); this is how it is usually presented.

1.4.3 Spectral signal representation

Spectral analysis contains very useful information about the speech signal. It is possible to calculate the spectrum of the entire sound or of part of it, and so to obtain a close realization of the static and dynamic spectral analysis of the sound signal. In a static analysis we assume a case where the spectrum is calculated from the entire sound

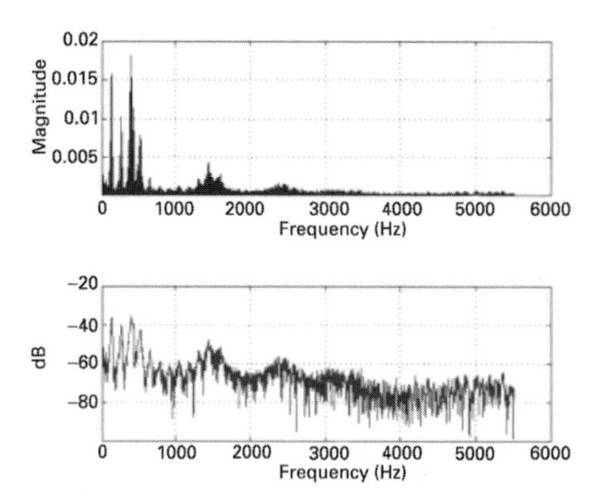

Figure 1.9 Spectrum of the word 'test'.

fragment from the beginning to the end. In the case of a dynamic spectral analysis we deal with a spectral calculation for a window of defined size, thus using only part of the file. This window is moving along the file. Hence, we can trace a frequency that changes in amplitude in the time flow.

If we need to calculate the spectrum of the sound file, we execute a static analysis. For this, we take the FFT for the entire sound file from the beginning to the end. Hence, the spectrum of the entire signal is determined, as shown in Figure 1.9 for the word 'test'. Looking at the waveform in Figure 1.8, we can observe the dependence of the amplitude on the time, but it is very difficult to estimate the frequency structure of the sound. In Figure 1.9, however, it is easy to see which frequencies dominate in the signal; we can see the noise level distinctly. It is also possible to say that the main frequencies of the vocal tract are in the range 80 Hz to 2 kHz and that some frequencies are very pronounced. However, the time dependence has been lost; in other words it is very difficult to estimate the variation in the time-dependent characteristics. To obtain a frequency–time analysis we can use *spectrograms*.

1.5 Speech coding techniques

In general, speech coding can be considered to be a particular speciality in the broader field of speech processing, which also includes speech analysis and speech recognition. The entire field of speech processing is currently experiencing a revolution that has been brought about by the maturation of digital signal processing (DSP) techniques and systems. Conversely, it is fair to say that speech processing has been a cradle for DSP, in the sense that many of the most widely used algorithms in DSP were developed or first brought into practice by people working on speech processing systems.

The purpose of a speech coder is to convert an analogue speech signal into digital form for efficient transmission or storage and to convert a received digital signal back to analogue [28].

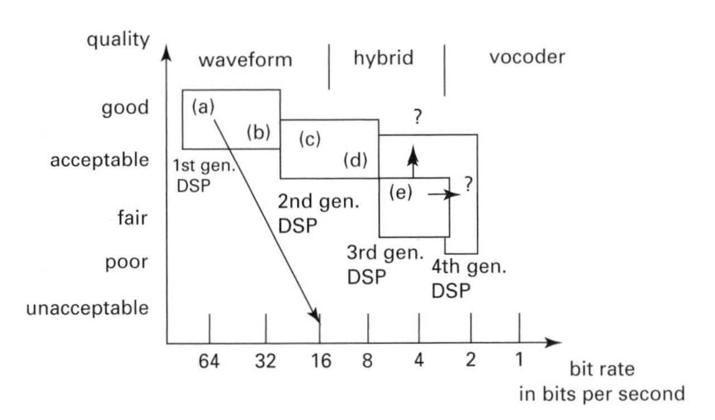

Figure 1.10 Speech coding quality versus bit rate. The diagonal arrow shows that first-generation DSP could require a bit rate as low as 16 kbps [29].

The main thrust of recent work on speech coding has been to reduce the transmission bit rate. Just a few years ago, the prospect that low-bit-rate speech coders would be used for public telecommunications services seemed remote; there was little demand and the available speech quality was too poor. The development in speech coding techniques in the last decade has been very rapid, and greatly improved speech quality at low bit rates has been achieved [3].

Transmission errors are inevitable when the services of digital radio links are used. For example, the capacity of digital cellular telephone systems is restricted by radio interference from nearby cells using the same frequencies, which produces digital errors in the received signal. Low-bit-rate speech coders are utilized for these services to save bandwidth and also because they can be made very robust to digital errors. Figure 1.10 shows the trade-off between bit rate and quality for some encoding techniques.

Speech quality and bit rate are two factors that directly conflict with each other. For systems that connect to the Public Switched Telephone Network (PSTN) and associated systems, the quality requirements are strict and must conform to constraints and guidelines imposed by the relevant regulatory bodies, e.g. the CCITT (International Consultative Committee for Telephone and Telegraph).

The quality objective for speech coders utilized in public telecommunications systems is generally to achieve the same speech quality as in a long-distance telephone network (PSTN), frequently referred to as 'toll quality'.

Speech coding algorithms can be classified into three types:

- waveform coding
- vocoding
- hybrid coding

We will consider these in the next subsection.

1.5.1 Waveform coding

Waveform coders can be divided into two types, static and dynamic. The first type does not adjust to changing speech input and therefore is very simple. The second type is the dynamic coder, which compensates for variations in amplitude of the speech signal by changing the quantizer step size. This change can happen over a few samples or a long period of time. The dynamic changes can come in the form of feedforward or feedback adaptation. Feedforward adaptation produces compensations based on the input to the coder, hence requiring extra information to be sent to the receiver and reducing the bit rate. A feedback system uses the coder output to produce a dynamic change. Predictor coefficients and reconstruction levels are calculated using the coder output. There is no need to send any additional information since the encoder and decoder both have access to the coded signal.

Waveform coders try to reconstruct the input signal's waveform. They are generally designed to be signal independent so they can be used to code a wide range of signals. They also offer a good degradation in the noise and transmission errors. Nevertheless, to be effective they can only be used for medium bit rates. Waveform coding can be done in either the time or the frequency domains [28–30].

Time domain coding

Pulse code modulation (PCM) The PCM algorithm is the most familiar and widely used waveform coder. Simulating the amplitude versus time waveform on a sample-to-sample basis approximates the analogue input waveform [4].

The incoming speech waveform is sampled and then logarithmically quantized into discrete levels. There are two methods of log PCM, which produce bit rates of 64 kbps; these are the A-law and μ-law methods [28]. In these methods a larger quantization step size is used for large-amplitude signals and a smaller step size for small-amplitude signals. With this type of coding, 13-bit samples can be compressed to 8 bits while keeping the original quality of speech. A-law PCM is the standard for PCM in Europe ($A = 87.56$) while the United States has used μ-law PCM as its standard ($\mu = 255$) [2].

PCM is the simplest type of waveform coding. It is basically just a quantization process. Every sample in the coder is quantized to a finite number of reconstruction levels, and to each level is assigned a unique sequence of binary digits; it is this sequence that is transmitted to the receiver. Scalar quantization can be used with this scheme, but the most common form of quantization used is logarithmic quantization. An 8 bit A-law or μ-law PCM is defined as the standard method of coding telephone speech. These schemes are very popular because of their low complexity and encoding delay [1, 2].

Differential pulse code modulation (DPCM) The waveform in PCM is coded without any assumptions about its nature and therefore PCM works very well for non-speech signals. There is a very high correlation between adjacent samples in the speech coding. This feature is employed to reduce the resulting bit rate [2]. This can be done by transmitting only the differences between each sample. This technique is known as differential pulse code modulation (DPCM) [1, 2]. The transmitted signal (the difference

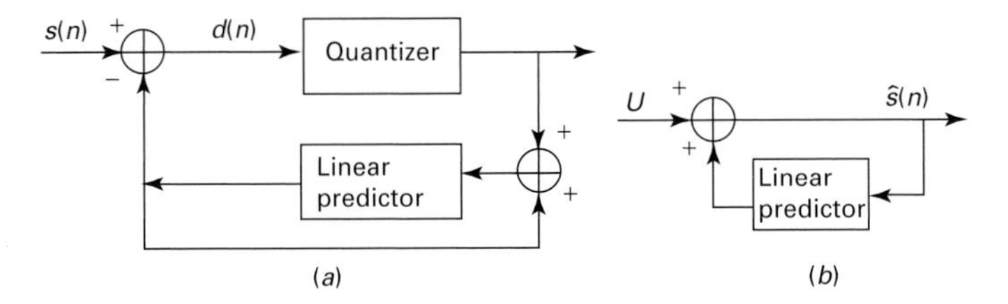

Figure 1.11 A DPCM system: (*a*) the coder and (*b*) the decoder; u is the excitation signal and $\hat{s}$ is the predicted signal [4].

signal) will have a much lower dynamic range than the original speech; hence it can be effectively quantized using a quantizer with fewer reconstruction levels. The previous sample is used to predict the value of the present sample. The prediction is improved if a much larger block of the speech is used to make the prediction.

A block diagram of a DPCM system is shown in Figure 1.11. Usually the predicted value, $\hat{s}(n)$, is a linear combination of a finite number of past speech samples $s(n-k)$:

$$\hat{s}(n) = \sum_{k=1}^{p} a_k s(n-k) \tag{1.2}$$

The difference signal,

$$d(n) = s(n) - \hat{s}(n) \tag{1.3}$$

is known as the residual and it is this residual that is quantized and transmitted to the receiver. The predictor's coefficients, a_k, are selected to minimize the mean square prediction error, E, which is given by

$$E = \sum_{n} [s(n) - \hat{s}(n)]^2 \, . \tag{1.4}$$

Delta modulation (DM) Delta modulation is a subclass of DPCM in which the difference (the prediction error E) is encoded with only one bit. Delta modulation typically operates at sampling rates much higher than the rates commonly used with DPCM. The step size in DM may also be adaptive (ADM). The DM and DPCM are low-to-medium-complexity coders and perform better than ordinary PCM for rates at and below 32 kbps [4].

Adaptive differential pulse code modulation (ADPCM) With DPCM both the predictor and the quantizer remain fixed in time. Greater efficiency can be achieved, however, if the quantizer is *adapted* to the changing statistics of the prediction residual. Further gains can be made if the predictor itself can adapt to the speech signal. This ensures that the mean square prediction error is being continually minimized independently of the speaker and the speech signal [31, 32]. Figure 1.12 shows an ADPCM system with both types of adaptation.

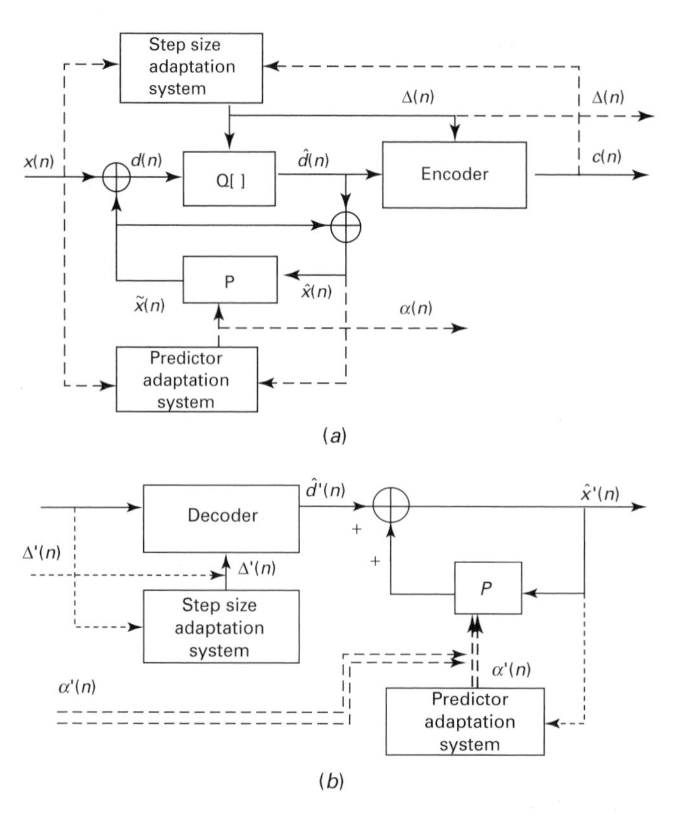

Figure 1.12 ADPCM system with both adaptive quantization and adaptive prediction: (*a*) coder; (*b*) decoder [1].

There are two methods for adapting quantizers and predictors, namely feedforward and feedbackward adaptation, mentioned earlier. With feedforward adaptation the reconstruction levels and the prediction coefficients are calculated at the transmitter, using a block of speech. Then they are themselves quantized and transmitted to the receiver as side information. Both the transmitter and the receiver use these quantized values to make predictions and quantize the residual. For feedbackward adaptation the reconstruction levels and predictor coefficients are calculated using the coded signal. Since this signal is known to both transmitter and receiver, there is no need to transmit any side information, so the predictor and quantizer can be updated for every sample. Feedbackward adaptation can produce lower bit rates but it is more sensitive to transmission errors than feedforward adaptation.

Adaptive differential pulse code modulation is very useful for coding speech at medium bit rates. The CCITT has formalized a standard for coding telephone speech in 32 kbps. This is the G.721 standard [33]. It uses a feedbackward adaptation scheme for both the quantizer and the predictor, as shown in Figure 1.12.

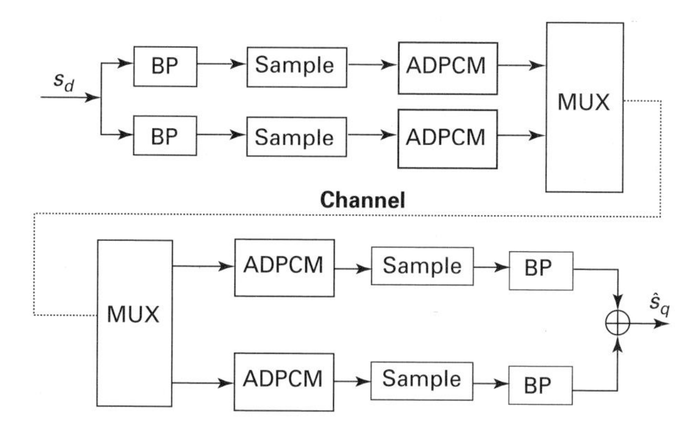

Figure 1.13 A subband coder (two bands shown); MUX is a multiplexer [34].

Frequency domain coding

Frequency domain waveform coders divide the signal into a number of separate frequency components and encode these independently. The number of bits used to code each frequency component can be varied dynamically.

Subband coding This is the simplest of the frequency domain techniques [34]. The signal in a subband coder is passed through a bank of bandpass filters, as shown in Figure 1.13. The output of each filter is then sampled. One of the time domain techniques described above is used to code the subbands. Each band has a number of bits assigned, which can be varied according to the band's perceptual importance. The sampling rates at the receiver are increased and the bands are modulated back to their original locations. Then they are summed to give the output speech.

The most important advantage of subband coding is that the quantization noise produced in one band is restricted to that band. This prevents the quantization noise from masking frequency components in other bands and allows split quantizer step sizes to be used for each band. Thus bands with lower energy can have lower step sizes and hence are conserved in the reconstructed signal. The presence of the quantization noise also permits the perceptually weighted distribution of bits.

Transform coding This is a more complex technique and involves the block transformation of a windowed segment of the input signal [35]. The idea is that the signal is transformed into the frequency, or some other similar, domain. Coding is then accomplished by assigning more bits to the more important transform coefficients. At the receiver the decoder then carries out the inverse transform to obtain the reconstructed signal.

Several transforms, such as the discrete Fourier transform (DFT) or the Karhunen–Loeve transform (KLT), can be used [36]. The KLT is the optimum transform and will decorrelate the signal. The most commonly used transform, however, is the discrete cosine transform (DCT), which is defined as follows:

$$X_c(k) = \sum_{n=0}^{N-1} x(n)g(k)\cos\left[\frac{(2n+1)k\pi}{2N}\right], \qquad k = 0, 1, \ldots, N-1 \tag{1.5}$$

where

$$x(n) = \frac{1}{N}\sum_{k=0}^{N-1} X_c(k)g(k)\cos\left[\frac{(2n+1)k\pi}{2N}\right], \qquad n = 0, 1, \ldots, N-1 \tag{1.6}$$

with $g(0) = 1$ and $g(k) = \sqrt{2}, k = 1, \ldots, N-1$. The DCT is used because it is significantly less computationally intense than the KLT and its properties are almost the same.

Transform coding is widely used in coding wideband audio and image signals. However, it is not normally used in coding speech signals because of its complexity.

1.5.2 Vocoding

The model for vocoders (VOice CODERS) assumes that speech is created by exciting a linear system, the vocal tract, by a sequence of periodic pulses if the sound is voiced or by noise if it is unvoiced [25].

If the speech is voiced the excitation consists of a periodic series of impulses; the distance between these pulses equals the pitch period. If the speech is unvoiced then the excitation is a random noise sequence, corresponding to the hiss produced by air blowing through a constriction in the vocal tract. The speech production model used in vocoders was shown in Figure 1.4.

A linear system models the vocal tract and its parameters can be determined using several techniques. Vocoders attempt to produce a signal that sounds like the original speech. At the transmitter the speech is analysed to determine the model parameters and the excitation. This information is then transmitted to the receiver where the speech is synthesized. The result of this is that vocoders can produce intelligible speech at very low bit rates. However, the synthesized speech sounds unnatural; hence vocoders are normally used where a low bit rate is of the maximum importance.

The poor quality of the vocoder output is attributable to the very simple nature of its speech production model, especially the assumption that speech is either voiced or unvoiced, allowing for no intermediate states. The ear is very sensitive to pitch information, so for voiced speech the pitch must be accurately determined, a problem that has never been satisfactorily solved. Vocoders also suffer from sensitivity to errors in the vocal tract model, errors occurring in either calculating the model's parameters or transmitting the data to the receiver [37].

Channel vocoder The ear's insensitivity to short-time phase is used to advantage in designing this coder. The magnitude of the spectrum for speech segments of about

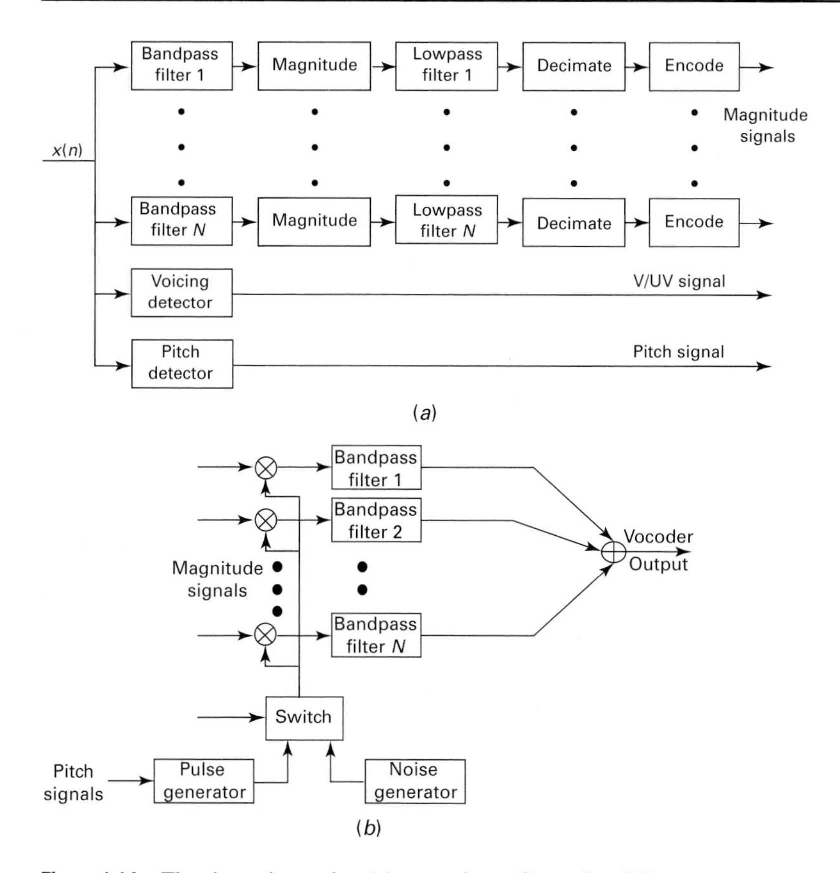

Figure 1.14 The channel vocoder: (*a*) transmitter; (*b*) receiver [1].

20 ms needs to be taken into account. A filter bank is used for estimating the spectrum. A better estimation can be achieved with more filters in this bank, but the bit rate will be higher. The output of each of these filters is then rectified and low-pass filtered to find the envelope of the signal. It is then sampled and transmitted to the receiver [1]; the exact opposite is done at the transmitter corresponding to the receiver. This is shown in Figures 1.14(*a*) and (*b*) respectively.

The bandwidths of the filters used in the filter bank tend to increase with frequency. This allows for the fact that the human ear responds linearly to a logarithmic scale of frequencies. The channel vocoder can provide highly intelligible speech at bit rates in the region of 2.4 kbps [38].

Homomorphic vocoder The basic idea in homomorphic vocoders is that the vocal-tract and the excitation log-magnitude spectra can be combined additively to produce the speech log-magnitude spectrum [1]:

$$\log\left(|S(e^{i\omega})|\right) = \log(|P(e^{i\omega})|) + \log(|V(e^{i\omega})|) \tag{1.7}$$

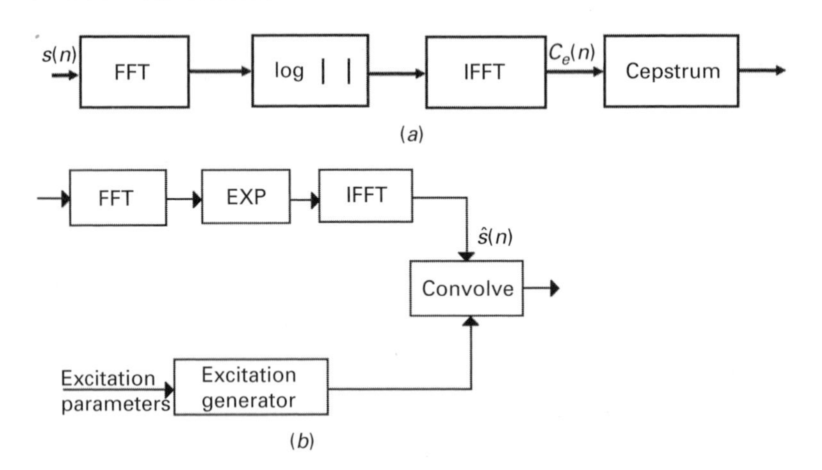

Figure 1.15 A homomorphic vocoder: (*a*) analysis; (*b*) synthesis.

where $S(e^{i\omega})$ is the spectrum of the speech, $P(e^{i\omega})$ is the spectrum of the excitation and $V(e^{i\omega})$ is the vocal tract spectrum.

A speech analysis-by-synthesis system, the homomorphic vocoder, that uses the cepstrum is shown in Figure 1.15. The inverse Fourier transform of the log-magnitude spectrum of the speech signal produces a cepstral sequence $C_e(n)$. The frequency samples of the cepstrum, which are near the origin, are associated with the vocal tract [7]. The coefficients $C_e(n)$ can be extracted using a cepstral window. The length of the cepstral window must generally be shorter than the shortest possible pitch period. For voiced speech the cepstral sequence has large samples at the pitch period. Therefore, the fundamental frequency can be estimated from the cepstrum [7]. The synthesizer takes the FFT of the cepstrum and the resulting frequency components are exponentiated. The IFFT of these components gives the impulse response of the vocal tract, which is convolved with the excitation to produce synthetic speech.

Although the cepstral vocoder did not find many applications at the time it was proposed, cepstrum-based methods for pitch and vocal tract estimation have found many other speech processing applications. In addition, in a fairly recent contribution by Chung and Schafer [39] it was reported that good-quality speech is obtainable at 4.8 kbps.

Formant vocoder The information in a speech signal is concentrated in the positions and bandwidths of the vocal tract's formant frequencies. Therefore if these formants can be accurately determined then it is possible to obtain a very low bit rate. Actually, with this technique it is possible to achieve bit rates less than 1 kbps. Nevertheless, the formants are very difficult to determine precisely. Therefore the formant vocoder has never been particularly preferred [40].

Linear predictive vocoder The same speech production model is used as for other vocoders. The difference is just in the method used to determine the model of the vocal tract. For this coder it is supposed that an all-pole infinite-impulse response filter, $H(z)$, can describe the vocal tract [2, 30]. Each speech sample is assumed to be a linear combination of the previous samples. The coefficients of this filter are calculated to minimize the error between the prediction and the actual sample.

A block of about 20 ms of speech is stored and analysed to calculate the predictor coefficients. These coefficients are then quantized and transmitted to the receiver. The speech is now passed through the inverse of the vocal tract filter to obtain the prediction error or residual. The effect of the predictor is to remove the correlation between adjacent samples. The predictor makes determining the pitch period of the speech much easier, by making the long-term correlation more visible. Hence a more reliable distinction between voiced and unvoiced waveforms can be made using this residual.

Linear predictive coders are the most popular of the vocoders because the all-pole model of the vocal tract works very well. They can be used to achieve highly intelligible speech at a bit rate as low as 2.4 kbps. The linear predictive model can also be used as the first stage in a formant vocoder, since the formants in the speech will closely correspond to the poles of the vocal tract filter [41].

1.5.3 Hybrid coding

A combination of the previous two methods (subsections 1.5.1 and 1.5.2) is called hybrid coding. Waveform coders try to maintain the waveform of the signal to be coded. They can provide very high quality speech at medium bit rates (32 kbps), but they cannot be used to code speech at very low bit rates. Vocoders, however, try to produce the correct signal, i.e. one which is like the input. They can be used to obtain very low bit rates but the reconstructed speech sounds very synthetic. It is perhaps specious to suppose that by combining the two techniques in some way a high-quality speech coder operating at low bit rates (less than 8 kbps) could be produced. In fact, in a hybrid coder one would merely attempt to preserve the perceptually important parts of the input speech waveform [3].

Residual excited linear prediction The residual excitation signal essentially carries all the information that has not been captured by linear prediction analysis, i.e. phase and pitch information, zeros due to nasal sounds etc. Although the concept of coding the prediction residual is also utilized in ADPCM and in adaptive predictive coders, RELP is different in that the residual encoding is based on spectral rather than waveform matching. In addition, RELP coders rely on the fact that the low-frequency components of speech are perceptually important [21].

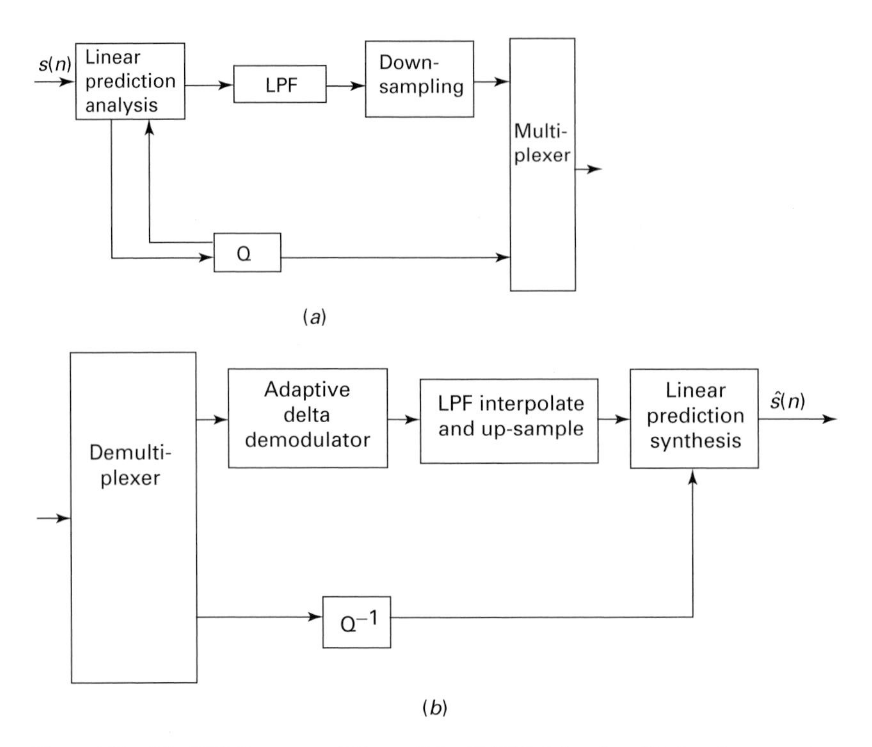

Figure 1.16 The RELP system of Un and Magill [42]: (*a*) coder; (*b*) decoder.

A RELP coder operating between 6 and 9.6 kbps was proposed in the mid-1970s by Un and Magill [42]. A block diagram for this RELP coder and decoder is shown in Figure 1.16. The RELP coder compresses the bandwidth of the residual to 800 Hz, thereby coding only the baseband of the residual at 5 kbps. In the decoder, the residual is down-sampled and coded using ADM techniques. At the receiver the baseband residual is processed by a non-linear spectral flattener whose function is to regenerate the high-frequency harmonics. The excitation of the synthesis filter is achieved by combining the flattened residual with an appropriate amount of white random noise.

High-frequency regeneration can also be achieved by operating directly on the frequency components of the residual.

A block diagram for a RELP coder and decoder that encodes the residual in the frequency domain using a fast Fourier transform (FFT) is shown in Figure 1.17. In this system, the FFT of the residual is computed and the magnitudes and phases of the frequency components within the baseband (typically below 1 kHz) are encoded and transmitted. At the receiver, a pitch-dependent high-frequency procedure is performed to generate the high-frequency residual.

A RELP coder that employs long-term prediction and adaptive bit allocation was proposed by Fette [43]. This coder was also one of the candidates for the 4.8 kbps Federal Standard 1016. A RELP coder that uses vector quantization for the encoding of the residual and linear prediction parameters was proposed by Adoul *et al.* [44].

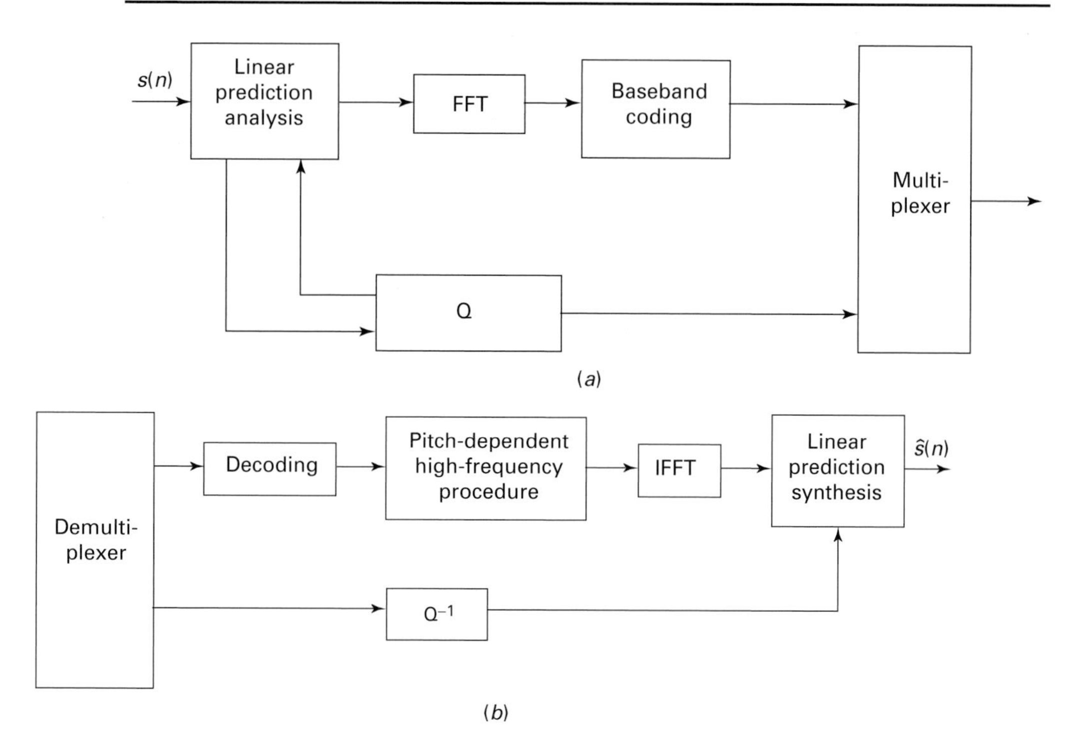

Figure 1.17 The FFT-based RELP system [42]: (*a*) coder; (*b*) decoder.

However, RELP is limited by the information lost in the residual baseband filtering. The analysis-by-synthesis linear predictive coders presented in the next chapter avoid this problem by using efficient excitation models that can be optimized both for waveform matching and perception.

Multipulse coding (MPC) The major problem with the model of the vocal tract used in the linear predictive vocoder is the excitation of the vocal tract. The vocoder classifies the speech as either voiced or unvoiced; there is nothing in between. This is the reason for the synthetic-sounding output. Multipulse excitation attempts to rectify this problem.

The correlation between adjacent samples is removed when the speech is passed through the linear predictor. The pitch of the speech for voiced speech introduces a long-term correlation into the speech, resulting in quasi-periodicity. The periodicity is not removed by the linear predictor and produces large spikes in the residual. This long-term correlation can be removed by passing the residual through a second linear predictor (a pitch predictor). This second predictor is designed not to remove the correlation from adjacent samples but to remove the correlation from adjacent periods of the residual. This is achieved by inserting a delay corresponding to the pitch period into the predictor, so that it is of the following form:

$$P(z) = 1 - \sum_i B_i z^{\tau - i} \tag{1.8}$$

where τ is the pitch period and B_i is the pitch gain. The two linear predictors are excited with a series of impulses. Normally about four to six impulses are used for the excitation. The location and amplitudes of these impulses are determined using an analysis-by-synthesis (AbS) procedure.

The multipulse coder is capable of producing high-quality speech at a bit rate around 9.6 kbps or lower. A modification of the multi-pulse coder is the regular-pulse-excitation (RPE) coder. The RPE coder uses regularly spaced pulse patterns instead of the separate impulses of the multipulse coder. The GSM standard uses an RPE coder operating at 13 kbps [19].

Code-excited linear prediction (CELP) In the CELP coder, the speech is passed through a cascade formed by the vocal tract predictor and the pitch predictor. The output of this predictor is a good approximation to Gaussian noise. This noise is quantized and transmitted to the receiver. Vector quantization is used in CELP coders. The index of the codeword that produces the best quality speech is transmitted along with an associated gain [37].

The codebook search is carried out using an analysis-by-synthesis technique. The speech is synthesized for every entry in the codebook. The codeword that produces the lowest error is chosen as the excitation. The error measure used is perceptually weighted so that the chosen codeword produces the speech that sounds the best.

A CELP coder can now be implemented in real time using modern digital-signal-processing microprocessors. This technique is currently one of the most effective methods of obtaining high-quality speech at very low bit rates. For example, the US Federal Standard 1016 describes a CELP coder that operates at 4.8 kbps for narrowband or telephone speech [45].

Sinusoidal modelling The sinusoidal transform coder (STC) resembles the transform coders mentioned in subsection 1.5.1 [13, 14]. These coders depict the speech signal as a linear combination of L sinusoids with time-varying amplitudes, phases and frequencies. Voiced speech can be represented by a small number of sinusoids because it tends to be highly periodic. Unvoiced speech can also be represented by a small number of sinusoids but with appropriately random phases. This allows the speech signal to be described using just a few parameters and consequently produces a good compression ratio. The sinusoids can be calculated from the short-time Fourier transform (STFT). The sinusoids selected are those that correspond to the peaks of the STFT. The speech is windowed using a Hamming window of length 2.5 times the average pitch. This is essential to provide enough resolution in the frequency domain.

The algorithm is not very dependent on the pitch frequency; it is only used to determine the window length. For this reason the STC is one of the best speech coders for non-speech signals. For very-low-bit-rate applications the frequencies of the sinusoids can be constrained to be integer multiples of the pitch frequency.

Multiband excitation (MBE) The motivation for multiband excitation (MBE) is to achieve higher quality speech than that produced by vocoders, while avoiding the computational cost of CELP [15]. The MBE coder uses the same procedure (i.e. speech is produced by exciting a vocal tract model) as traditional vocoders. The vocal tract model is the same as that used in vocoders but the method of modelling the excitation is different. With traditional vocoders the excitation is defined by a voiced/unvoiced decision. The excitation is a noise sequence for unvoiced speech and a periodic impulse train for voiced speech. The MBE coder replaces the single voiced/unvoiced decision with a series of such decisions. The excitation is divided into several subbands, and in each subband the spectrum of the speech is analysed to determine whether it is voiced or unvoiced. The result is a series of voiced/unvoiced decisions. As with vocoders, the voiced segments are coded using an impulse train and the unvoiced segments using noise. This allows the coded speech to display voiced and unvoiced characteristics simultaneously, just as real speech does.

An improved MBE (IMBE) coder was proposed by Hardwick and Lim [22]. Although the IMBE coder is based on the MBE analysis-by-synthesis model, it employs more efficient methods for quantizing the MBE model parameters. In addition, the IMBE coding scheme is more robust to channel impairment [22].

REFERENCES

[1] L. R. Rabinar and R. W. Schafer. *Digital Processing of Speech Signals*. Prentice Hall, 1978.

[2] N. S. Jayant and P. Noll. *Digital Coding of Waveforms: Principles and Applications to Speech and Video*. Prentice Hall, 1984.

[3] F. A. Westall and S. F. A. Ip. *Digital Signal Processing in Telecommunications*. Chapman and Hall, 1993.

[4] N. S. Jayant. Digital coding of speech waveforms: pcm, dpcm, and dm quantisers. *Proc. IEEE*, **62**, 611–32, 1974.

[5] J. Makhoul. Linear prediction analysis. *Proc. IEEE*, **63**(4), 561–80, 1974.

[6] B. Atal and S. Hanauer. Speech analysis and synthesis by linear prediction of the speech waves. *Proc. IEEE*, **50**, 637, 1971.

[7] D. G. Childers. The cepstrum. *Proc. IEEE*, **10**, 1428–43, 1977.

[8] R. Schafer and L. Rabiner. Design and simulation of a speech analysis–synthesis system based on short-time Fourier analysis. *Trans. Audio and Electroacoustics*, **AU21**(3): 165, 1973.

[9] M. R. Portnoff. Time–frequency representation of digital signals and systems based on short-time Fourier analysis. *IEEE Trans. ASSP*, **28**(1): 55–69, 1980.

[10] M. R. Portnoff. Implementation of the digital phase vocoder using the Fast Fourier Transform. *IEEE Trans. ASSP*, **24**(3): 243, 1976.

[11] J. Tribolet and R. Crochier. Frequency domain coding of speech. *IEEE Trans. ASSP*, **27**(5): 512, 1979.

[12] T. E. Termain. The government standard linear predictive coding algorithm: lpc-10. In *Speech Technology*, pp. 40–9, 1982.

[13] R. McAulay and T. Quatieri. Speech analysis/synthesis based on a sinusoidal representation. *IEEE Trans. ASSP*, **34**(4): 744, 1986.

[14] R. McAulay and T. Quatieri. Multirate sinusoidal transform coding at rates from 2.4 kbps to 8 kbps. In *Proc. ICASSP87, Dallas*, 1987.

[15] D. Griffin and J. Lim. Multiband excitation vocoder. *IEEE Trans. ASSP*, **36**(8): 1223, 1988.

[16] B. Atal and J. Remde. A new model for lpc excitation for producing natural sounding speech at low bit rates. In *Proc. ICASSP82*, pp. 614–17, 1982.

[17] I. Gerson and M. Jasiuk. Vector sum excited linear prediction (vselp) speech coding at 8 kbps. In *Proc. ICASSP90, New Mexico*, pp. 461–4, 1990.

[18] M. R. Schroeder and B. Atal. Code excited linear prediction (celp): high quality speech at very low bit rates. In *Proc. ICASSP90, Tampa*, p. 937, 1990.

[19] P. Vary *et al.* Speech coder for the European mobile radio system. In *Proc. ICASSP88, Tampa*, p. 227, 1988.

[20] D. Kemp *et al.* An evaluation of 4800 bits/s voice coders. In *Proc. ICASSP89, Glasgow*, p. 200, 1989.

[21] B. Atal, V. Cuperman, and A. Gersho. *Advances in Speech Coding*. Kluwer Academic, 1990.

[22] J. Hardwick and J. Lim. The application of the imbe speech coder to mobile communications. In *Proc. ICASSP91*, pp. 249–52, 1991.

[23] S.-W. Wong. An evaluation of 6.4 kbps speech coders for the inmarsatm system. In *Proc. ICASSP91, Toronto*, pp. 629–32, 1991.

[24] J. L. Flanagan. *Speech Analysis Synthesis and Perception*. Springer, 1972.

[25] T. Parsons. *Voice and Speech Processing*. McGraw-Hill, 1987.

[26] A. M. Kondoz. *Digital Speech Coding for Low Rate Communications Systems*. John Wiley and Sons, 1994.

[27] A. V. Oppenheim. *Applications of Digital Signal Processing*. Prentice-Hall, 1978.

[28] R. Steel. *Mobile Radio Communications*. Prentice-Hall, 1992.

[29] F. J. Owens. *Signal Processing of Speech*. Macmillan, 1993.

[30] S. Saito and K. Nakata. *Fundamentals of Speech Signal Processing*. Academic Press, 1987.

[31] J. Gibson. Adaptive prediction for speech encoding. *ASSP Mag.*, **1**: 12–26, 1984.

[32] J. Gibson. Sequentially adaptive backward prediction in adpcm speech coders. *IEEE Trans. Comm.*, pp. 145–50, 1978.

[33] CCITT Recommendation G.721. 32 kbps adaptive differential pulse code modulation (adpcm). *Blue Book*, **III**(III): 145–50, 1988.

[34] R. V. Cox *et al.* New directions in subband coding. *IEEE Trans. Selected Areas in Communications: Special Issue on Voice Coding for Communications*, **6**(2): 391–409, 1988.

[35] R. V. Cox and R. Crochiere. Realtime simulation of adaptive transform coding. *IEEE Trans. ASSP*, **29**(2): 147, 1981.

[36] R. Zelinski and P. Noll. Adaptive transform coding of speech signals. *IEEE Trans. ASSP*, **29**: 299, 1977.

[37] M. R. Schroeder. Vocoders: analysis and synthesis of speech. In *Proc. IEEE*, **54**, 720–34, 1966.

[38] B. Gold *et al.* New applications of channel vocoders. *IEEE Trans. ASSP*, **29**: 13, 1981.

[39] J. Chung and R. Schafer. Excitation modeling in a homomorphic vocoder. In *Proc. ICASSP90, New Mexico*, p. 25, 1990.

[40] J. N. Holmes. Formant synthesizer: cascade or parallel? In *Proc. Speech Communication, North Holland*, pp. 251–73, 1983.

[41] V. Cuperman and A. Gersho. Vector predictive coding of speech at 16 kbps. *IEEE Trans. Com*, **33**: 685, 1985.

[42] C. K. Un and D. T. Magill. The residual excited linear prediction vocoder with transmission rate below 9.6 kbps. *IEEE Trans. Com*, **23**(12): 1466, 1975.

[43] B. Fette. Experiments with high quality low complexity 4800 bps residual excited lpc (relp). In *Proc. ICASSP88, New York*, pp. 263–6, 1988.

[44] J. Adoul *et al.* 4800 bps relp vocoder using vector quantisation for both filter and residual representation. In *Proc. ICASSP82*, p. 601, 1982.

[45] J. Campbell, T. E. Tremain and V. Welch. The proposed federal standard 1016 4800 bps voice coder: celp. In *Speech Technology*, pp. 58–64, 1990.

2 Computational background

The computational background to digital signal processing (DSP) involves a number of techniques of numerical analysis. Those techniques which are of particular value are:

- solutions to linear systems of equations
- finite difference analysis
- numerical integration

A large number of DSP algorithms can be written in terms of a matrix equation or a set of matrix equations. Hence, computational methods in linear algebra are an important aspect of the subject. Many DSP algorithms can be classified in terms of a digital filter. Two important classes of digital filter are used in DSP, as follows.

Convolution filters are nonrecursive filters. They use linear processes that operate on the data directly.

Fourier filters operate on data obtained by computing the discrete Fourier transform of a signal. This is accomplished using the fast Fourier transform algorithm.

2.1 Digital filters

Digital filters fall into two main categories:

- real-space filters
- Fourier-space filters

Real-space filters Real-space filters are based on some form of 'moving window' principle. A sample of data from a given element of the signal is processed giving (typically) a single output value. The window is then moved on to the next element of the signal and the process repeated. A common real-space filter is the finite impulse response (FIR) filter. This is a non-recursive filter (in a non-recursive filter the output depends only on the present and previous inputs), involving a discrete convolution operation for sample i of the form

$$s_i = \sum_j p_{i-j} f_j \tag{2.1a}$$

where f_j is the input signal or object function, s_i is the output signal and p_{i-j} is the 'kernel' of the filter and is also called the *instrument function*. The subscript on p reads 'i minus j'.

Fourier-space filters Fourier-space filters usually involve multiplicative operations that act on the discrete Fourier transform (DFT) of the signal. If S_i, P_i and F_i are taken to denote the DFTs of s_i, p_i and f_i respectively then Eqn 2.1a transforms in Fourier space to

$$S_i = P_i F_i \tag{2.1b}$$

Equations 2.1a, b give two forms of the (discrete) convolution theorem. If p_i is composed of just a few elements then the discrete convolution Eqn 2.1a can be computed directly. If p_i is composed of many elements then it is numerically more efficient to use a fast Fourier transform (FFT) and perform the filtering operation in Fourier space.

Many problems associated with the processing of digital signals are related to what mathematicians generally refer to as *inverse problems*. Consider again Eqn 2.1a. This discrete convolution can be written in the form

$$\mathbf{s} = \mathbf{pf}$$

where $\mathbf{p}$ is a matrix whose elements are the components of the kernel p_i arranged in an appropriate order. A common inverse problem encountered in DSP is: 'given $\mathbf{s}$ and $\mathbf{p}$ compute $\mathbf{f}$'. This inverse problem is called *deconvolution*. Since a convolution can be written in terms of a matrix equation, the solution to this problem is expressed in terms of the solution to a set of linear simultaneous equations. If we express the same problem in Fourier space then we have

$$S_i = P_i F_i$$

The problem now becomes: 'given S_i and P_i find F_i'. In this case, the solution appears trivial since

$$F_i = \frac{S_i}{P_i}$$

$1/P_i$ is known as the *inverse filter*.

2.2 The fast Fourier transform

I thought it would be a good idea to remind the reader about a man whose contribution to science and engineering has been very great. Jean Baptiste Joseph, Baron de Fourier, was 30 years old when he took part in Napoleon's Egyptian campaign of 1798. He was made Governor of Lower Egypt, and contributed many scientific papers to the Egyptian Institute founded by Napoleon. He subsequently returned to France, and became Prefect of Grenoble.

Fourier submitted his ideas on the solution of heat flow problems using trigonometric series to the Institute de France in 1807. The work was considered controversial, and publication of his monumental book on heat had to wait another 15 years. He showed in his book that periodic signals can be represented as a weighted sum of harmonically related sinusoids. He also showed that non-repetitive, or non-periodic, signals can be considered as an integral of sinusoids that are not harmonically related. These two key ideas form the bases of the famous Fourier series and Fourier transform respectively and have had a profound influence on many branches of engineering and applied science, including electronic and signal processing.

The fast Fourier transform (FFT) is an algorithm for computing the discrete Fourier transform using fewer additions and multiplications. The standard-form DFT of an N-point vector is given for sample m by

$$F_m = \sum_{n=0}^{N-1} f_n e^{-2\pi i n m/N} \tag{2.2}$$

How much computation is involved in computing this N-point DFT? Write

$$W_N = e^{-2\pi i/N}$$

then

$$F_m = \sum_{n=0}^{N-1} W_N^{nm} f_n$$

This is a matrix equation, which can be written in the form

$$\begin{pmatrix} F_0 \\ F_1 \\ \vdots \\ F_{N-1} \end{pmatrix} = \begin{pmatrix} W_N^{00} & W_N^{01} & \cdots & W_N^{0(N-1)} \\ W_N^{10} & W_N^{11} & \cdots & W_N^{1(N-1)} \\ \vdots & \vdots & \ddots & \vdots \\ W_N^{(N-1)0} & W_N^{(N-1)1} & \cdots & W_N^{(N-1)(N-1)} \end{pmatrix} \begin{pmatrix} f_0 \\ f_1 \\ \vdots \\ f_{(N-1)} \end{pmatrix}$$

In this form, we see that the DFT is computed essentially by multiplying an N-point vector f_n by a matrix of coefficients given by a (complex) constant W_N to the power nm. This requires $N \times N$ multiplications. For example, to compute the DFT of 1000 points requires 10^6 multiplications!

Basic idea behind the FFT algorithm

By applying a simple but very elegant trick, an N-point DFT can be written in terms of two $N/2$-point DFTs. The FFT algorithm is based on repeating this trick again and again until a single-point DFT is obtained. Here is the trick: write

$$\begin{aligned}\sum_{n=0}^{N-1} f_n e^{-2\pi inm/N} &= \sum_{n=0}^{(N/2)-1} f_{2n} e^{-2\pi i(2n)m/N} + \sum_{n=0}^{(N/2)-1} f_{2n+1} e^{-2\pi i(2n+1)m/N} \\ &= \sum_{n=0}^{(N/2)-1} f_{2n} e^{-2\pi inm/(N/2)} + e^{-2\pi im/N} \sum_{n=0}^{(N/2)-1} f_{2n+1} e^{-2\pi inm/(N/2)} \\ &= \sum_{n=0}^{(N/2)-1} f_{2n} W_{N/2}^{nm} + W_N^m \sum_{n=0}^{(N/2)-1} f_{2n+1} W_{N/2}^{nm} \qquad (2.3)\end{aligned}$$

This gives us a fundamental result:

$$\begin{aligned}\text{DFT of } N\text{-point array } &= \text{DFT of even-numbered components} \\ &\quad + W_N^m \times \text{DFT of odd-numbered components}\end{aligned}$$

Using the superscripts e and o to represent 'odd' and 'even' respectively, we can write this result in the form

$$F_m = F_m^e + W_N^m F_m^o \qquad (2.4)$$

The important thing to note here is that the evaluation of F_m^e and of F_m^o is over $N/2$ points – the $N/2$ even-numbered components and the $N/2$ odd-numbered components of the original N-point array. To compute F_m^e and F_m^o we need only half the number of multiplications required to compute F_m.

Repeating the trick

Because the form of the expressions for F_m^e and F_m^o is identical to the form of the original N-point DFT, we can repeat the idea and decompose F_m^e and F_m^o into 'even' and 'odd' parts, producing a total of four $N/4$-point DFTs:

$$\begin{array}{c} F_m \\ \Downarrow \\ F_m^e \quad + \quad W_N^m F_m^o \\ \Downarrow \\ F_m^{ee} + W_{N/2}^m F_m^{eo} \quad + \quad W_N^m (F_m^{oe} + W_{N/2}^m F_m^{oo}) \end{array}$$

It is possible to continue subdividing the data into odd-numbered and even-numbered components until we get down to the DFT of a single point. Because the data is subdivided into 'odd' and 'even' components of equal length we require an initial array of size $N = 2^k, k = 1, 2, 3, \ldots$. Computing the DFT in this way reduces the number of multiplications needed to $N \log_2 N$, which, even for moderate values of N, is considerably smaller than N^2.

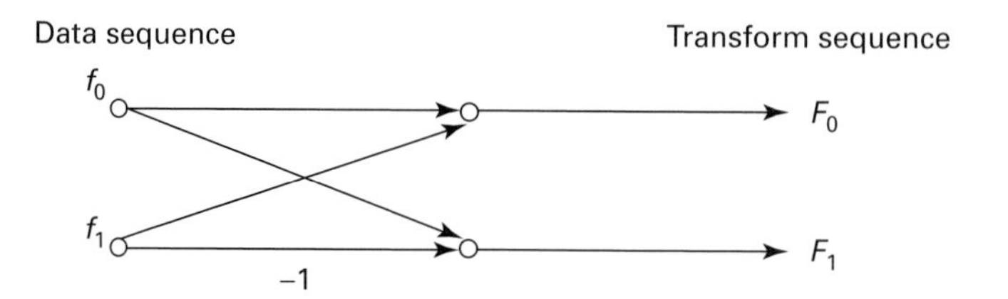

Figure 2.1 The trivial case: a two-point DFT. The input data are f_0 and f_1 and the corresponding outputs are F_0 and F_1.

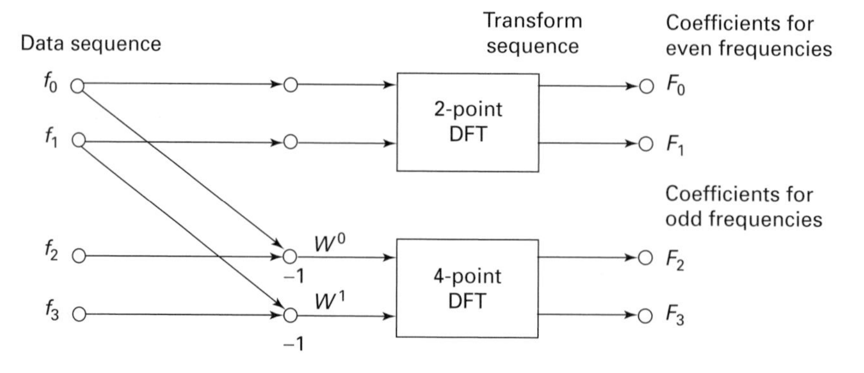

Figure 2.2 Reduction of a four-point DFT into two-point DFTs.

Examples

(i) Two-point FFT (f_0, f_1) This is shown in Figure 2.1. Using Eqn 2.3 we have

$$\begin{aligned}
F_m &= \sum_{n=0}^{1} f_n W_2^{nm} \\
&= W_1^0 f_0 + W_2^m W_1^0 f_1 \\
&= f_0 + e^{(i\pi m)} f_1 \\
F_0 &= f_0 + f_1 \\
F_1 &= f_0 + e^{(i\pi)} f_1 \\
&= f_0 - f_1
\end{aligned}$$

(ii) Four-point FFT (f_0, f_1, f_2, f_3) This is shown in Figure 2.2. We have

$$\begin{aligned}
F_m &= \sum_{n=0}^{3} f_n W_4^{nm} \\
&= \sum_{n=0}^{1} f_{2n} W_2^{nm} + W_4^m \sum_{n=0}^{1} f_{2n+1} W_2^{nm} \\
&= f_0 + W_2^m f_2 + W_4^m (f_1 + W_2^m f_3)
\end{aligned}$$

Substituting for m, we obtain

$$F_0 = f_0 + f_1 + f_2 + f_3$$
$$F_1 = f_0 + f_2 W_2 + f_1 W_4 + f_3 W_4 W_2$$
$$F_2 = f_0 + f_2 W_2^2 + f_1 W_4^2 + f_3 W_4^2 W_2^2$$
$$F_3 = f_0 + f_2 W_2^3 + f_1 W_4^3 + f_3 W_4^3 W_2^3$$

Implementation – bit reversal

Consider the eight-point array

$$f_0, f_1, f_2, f_3, f_4, f_5, f_6, f_7$$

The odd- and even-numbered elements, i.e. the elements with even and odd *arguments* (here, subscript labels) are displayed as follows, on the left and right respectively:

$$f_0, f_2, f_4, f_6 \qquad f_1, f_3, f_5, f_7$$

The even-numbered elements f_{2n} are now subdivided into an 'even' group for which n is even and an 'odd' group for which n is odd, and the odd-numbered elements f_{2n+1} are subdivided likewise.

even odd even odd

$$f_0, f_4 \quad f_2, f_6 \qquad f_1, f_5 \quad f_3, f_7$$

To use the FFT algorithm, the input array must be expressed in the form

$$f_0, f_4, f_2, f_6, f_1, f_5, f_3, f_7$$

The general procedure for re-ordering an input array follows a simple *bit-reversal* rule – the argument i of the element f_i in position i in the original array is expressed in binary form. The bits are then reversed to obtain the argument of the element that is now in position i in the re-ordered array, as shown in Table 2.1.

If the FFT algorithm is applied to an array in its natural order then the output is bit-reversed. Bit-reversal of the output is then required to obtain the correct order. Thus, bit-reversal can be applied either before or after the computations commence.

2.3 Data windowing

Computing the DFT of a digital signal with N samples is equivalent to multiplying an infinite run of sampled data by a 'window function' that is zero except during the total sampling time $N\Delta x$ and is unity during that time. Consider the following form of the

Table 2.1 *Bit-reversal for an eight-point transform*

Original argument	Original array	Bit-reversed argument	Re-ordered array
000	f_0	000	f_0
001	f_1	100	f_4
010	f_2	010	f_2
011	f_3	110	f_6
100	f_4	001	f_1
101	f_5	101	f_5
110	f_6	011	f_3
111	f_7	111	f_7

DFT, Eqn 2.2:

$$F_m = \sum_{n=-N/2}^{(N/2)-1} f_n \exp(-i2\pi nm/N) \tag{2.5}$$

When an N-point DFT is computed, the data are 'windowed' by a square window function. To identify the effect of computing an N-point DFT consider

$$F_m \sim \int_{-N/2}^{N/2} f(x)\exp(-ik_m x)\,\mathrm{d}x = \int_{-\infty}^{\infty} f(x)w(x)\exp(-ik_m x)\,\mathrm{d}x$$

where

$$w(x) = \begin{cases} 1, & |x| \leq N/2 \\ 0, & |x| > N/2 \end{cases}$$

Using the product theorem we can write (ignoring scaling constants)

$$F_m \sim \int_{-\infty}^{\infty} F(k_m - k)\frac{\sin(kN/2)}{kN/2}\,\mathrm{d}k$$

A sample of the discrete spectrum F_m obtained by taking the DFT of an N-point signal is given not by $F(k_m)$ but by $F(k_m)$ convolved with the 'sinc' function

$$\frac{\sin(kN/2)}{kN/2}$$

Note that

$$F_m \to F(k_m) \quad \text{as} \quad N \to \infty$$

Each sample F_m is an approximation to $F(k_m)$ that depends on the influence of the sinc function associated with one sample bin on the next sample bin: the sinc function 'leaks' from one bin to the next. The reason for this 'leakage' is that the square window turns on and off so rapidly that its Fourier transform has substantial components at high frequencies. To remedy this situation, we can multiply the input data f_n by a window function w_n that changes more gradually from zero to a maximum and then back to zero as n goes from $-N/2$ to $N/2$. Many windows exist for this purpose. The difference lies in the trade-off between the 'flatness' and 'peakedness' of the spectral leakage function' (i.e. the amplitude spectrum) of the window function.

Some commonly used windows

For sample i, $i = 0, 1, \ldots, N-1$, we have the following windows:

Parzen window

$$w_i = 1 - \left| \frac{i - \frac{1}{2}N}{\frac{1}{2}N} \right| \tag{2.6}$$

Welch window

$$w_i = 1 - \left(\frac{i - \frac{1}{2}N}{\frac{1}{2}N} \right)^2 \tag{2.7}$$

Hanning window (cosine taper)

$$w_i = \frac{1}{2}\left[1 - \cos\left(\frac{2\pi i}{N}\right)\right] \tag{2.8}$$

Hamming window

$$w_i = 0.54 - 0.46\cos\left(\frac{2\pi i}{N}\right) \tag{2.9}$$

Von Hann window (raised cosine)

$$w_i = \frac{1}{2}\left\{1 + \cos\left[\frac{\pi(i - \frac{1}{2}N)}{\frac{1}{2}N}\right]\right\} \tag{2.10}$$

Generalized von Hann window

$$w_i = b + 2a\cos\left[\frac{\pi(i - \frac{1}{2}N)}{\frac{1}{2}N}\right], \qquad 2a + b = 1 \tag{2.11}$$

Kaiser window

$$w_i = \frac{I_0\left\{\alpha\sqrt{1 - \left[\frac{(i - \frac{1}{2}N)}{\frac{1}{2}N}\right]^2}\right\}}{I} \tag{2.12}$$

where I_0 is the modified Bessel function of the first kind and I is a normalization coefficient.

2.4 Random number generation and noise

The performance of many DSP algorithms depends on the degree of noise present in the signal. Because many types of DSP algorithm are sensitive to noise, it is important to test their behaviour in the presence of noise. This is usually done by synthesizing noisy signals using number generators. Random numbers are not numbers generated by a random process but are numbers generated by a completely deterministic arithmetic process. The resulting set of numbers may have various statistical properties that together are regarded as displaying randomness.

A typical mechanism for generating random numbers is via the recursion

$$x_{n+1} = r x_n \mod P$$

where an initial element x_0 is repeatedly multiplied by r, each product being reduced modulo P. The element x_0 is commonly referred to as the *seed*.

Example With $r = 13$, $P = 100$ and $x_0 = 1$ we get the following sequence of two-digit numbers:

$$01, 13, 69, 97, 61, 93, 09, 17, 21, 73, 49, 37, 81, 53, 89, 57, 41, 33, 29, 77$$

For certain choices of r and P, the resulting sequence $x_0, x_1, x_2, \ldots$ is fairly evenly distributed over $(0, P)$, contains the expected number of upward and downward double runs (e.g. 13, 69, 97) and triple runs (e.g. 9, 17, 21, 73) and agrees with other predictions of probability theory. The values of r and P can vary. For example, with decimal computers $r = 7^9$ and $P = 10^s$ is a satisfactory choice.

The random number generator discussed in this section uses the Mersenne prime number $P = 2^{31} - 1 = 2147\,483\,648$ and $A = 7^5 = 16\,807$. Thus all integers x_n produced will satisfy $0 < x_n < 2^{31} - 1$. Before first use, x_0 – the seed – must be set to some initial integer value IX in the range $0 < IX < 2147\,483\,647$.

2.5 Computing with the FFT

Basic principle In general, the FFT can be used to implement digital algorithms derived from some theoretical result that is itself based on Fourier theory, provided that the data is adequately sampled to avoid aliasing and that appropriate windows are used to minimize spectral leakage.

2.5.1 Spectral analysis using the FFT

The FFT provides a complex spectrum, with real and imaginary arrays a_i and b_i respectively. From the output of the FFT we can construct a number of useful functions required for spectral analysis:

- the discrete amplitude spectrum $\sqrt{a_i^2 + b_i^2}$
- the discrete power spectrum $a_i^2 + b_i^2$
- the discrete (principal) phase spectrum $\tan^{-1}(b_i/a_i)$

The dynamic range of the amplitude and power spectra is often very low, especially at high frequencies. Analysis of the spectrum can be enhanced using the logarithmic function and generating a display of

$$\ln\left(1 + \sqrt{a_i^2 + b_i^2}\right)$$

or

$$\ln(1 + a_i^2 + b_i^2)$$

The term '1' is added in case $a_i^2 + b_i^2$ should equal 0.

Discrete convolution

To convolve two discrete arrays p_i and f_i of equal length using an FFT we use the convolution theorem

$$\begin{aligned} s_i &= p_i \otimes f_i \\ &\Downarrow \\ S_i &= P_i F_i \\ &\Downarrow \\ s_i &= \text{Re}[\text{IDFT}(S_i)] \end{aligned} \tag{2.13}$$

where S_i, P_i and F_i are the DFTs of s_i, p_i and f_i respectively, computed using the FFT, and where IDFT denotes the inverse DFT, also computed using the FFT. A typical algorithm will involve the following steps.

1. Input the real arrays p and f.
2. Set the imaginary parts associated with p, f and s (denoted by pp, ff and ss, say) to zero.
3. Call fft1d(f,ff,n,−1); call fft1d(p,pp,n,−1).
4. Do complex multiplication:

   s=p×f−pp×ff
   ss=p×ff+f×pp

5. Call fft1d(s,ss,n,1).
6. Write out s.

Using pseudo-code, the program for convolution using an FFT is as follows:

```
for i=1,2,...,n; do:
            fr(i)=signal1(i)
fi(i)=0.
pr(i)=signal2(i)
pi(i)=0.
fft(fr,fi)

fft(pr,pi)

for i=1,2,...,n; do:
sr(i)=fr(i)*pr(i)-fi(i)*pi(i)
si(i)=fr(i)*pi(i)+pr(i)*fi(i)

inverse
_fft(sr,si)
```

Discrete correlation

To correlate two real discrete arrays p_i and f_i of equal length using an FFT we use the correlation theorem

$$s_i = p_i \odot f_i$$

$$\Downarrow$$

$$S_i = P_i^* F_i$$

$$\Downarrow$$

$$s_i = \mathrm{Re}[\mathrm{IDFT}(S_i)] \tag{2.14}$$

where S_i, P_i and F_i are the DFTs of s_i, p_i and f_i respectively, computed using the FFT, and IDFT denotes the inverse DFT, also computed using the FFT. Correlation is an integral part of the process of convolution. The convolution process is essentially the correlation of two data sequences in which one of the sequences has been reversed. This means that the same algorithm may be used to compute correlations and convolutions simply by reversing one of the sequences [1].

A typical algorithm will involve the following steps.

1. Input the real arrays p and f.
2. Set the imaginary parts associated with p, f and s (denoted by pp, ff and ss, say) to zero.
3. Call fft1d(f,ff,n,−1); call fft1d(p,pp,n,−1).
4. Do complex multiplication:

   s=p×f+pp×ff
   ss=p×ff−f×pp

5. Call fft1d(s,ss,n,1).
6. Write out s.

Using pseudo-code, the algorithm for correlation using an FFT is:

```
for i=1,2,...,n; do:
          fr(i)=signal1(i)
fi(i)=0.
pr(i)=signal2(i)
pi(i)=0.
 fft(fr,fi)
fft(pr,pi)
 for i=1,2,...,n; do:
sr(i)=fr(i)*pr(i)+fi(i)*pi(i)
si(i)=fr(i)*pi(i)-pr(i)*fi(i)
 inverse_fft(sr,si)
```

2.5.2 Computing the analytic signal

The Argand diagram representation provides us with the complex representation of a signal:

$$s = A\exp(i\phi) = A\cos\phi + iA\sin\phi$$

where A is the amplitude and ϕ is the phase.

Let us write s in the form

$$s(x) = f(x) + iq(x)$$

where f is the real signal, q is the 'quadrature signal' and s is the 'analytic signal'.

Problem Given a real signal $f(x)$, how do we find the amplitude A and phase ϕ of the signal?

Solution First compute the imaginary part q – given by the *Hilbert transform* of f.

The Hilbert transform

Consider a real signal $f(x)$ with a complex spectrum $F(k)$; then

$$f(x) = \frac{1}{2\pi}\int_{-\infty}^{\infty} F(k)e^{ikx}\,\mathrm{d}k$$

Suppose we integrate over the physically meaningful part of the spectrum (i.e. the part with positive frequencies $k \geq 0$) and compute the function

$$s(x) = \frac{1}{\pi}\int_{0}^{\infty} F(k)e^{ikx}\,\mathrm{d}k$$

If $s(x)$, the analytic signal, is written as

$$s(x) = f(x) + iq(x)$$

then q, the Hilbert transform of f, is given by

$$q(x) = \frac{1}{\pi} \int_{-\infty}^{\infty} \frac{f(y)}{x - y} \, \mathrm{d}y \tag{2.15}$$

This result will now be proved.

Problem Show that

$$s(x) = \frac{1}{\pi} \int_{0}^{\infty} F(k) e^{ikx} \, \mathrm{d}k = f(x) + iq(x)$$

where $q(x)$ is the Hilbert transform of $f(x)$.

Solution Start by writing s in the form

$$s(x) = \frac{1}{2\pi} \int_{-\infty}^{\infty} 2U(k) F(k) \exp(ikx) \, \mathrm{d}k$$

where U is the step function given by

$$U(k) = \begin{cases} 1, & k \geq 0 \\ 0, & k < 0 \end{cases}$$

Using the convolution theorem, it follows that

$$s(x) = 2 \int_{-\infty}^{\infty} u(x - y) f(y) \, \mathrm{d}y$$

where

$$u(x) = \frac{1}{2} \delta(x) + \frac{i}{2\pi x}$$

and $\delta(x)$ is the delta function.

The analytic signal $s(x)$ can then be written in the form

$$\begin{aligned} s(x) &= \int_{-\infty}^{\infty} \left[\delta(x - y) + \frac{i}{\pi(x - y)} \right] f(y) \, \mathrm{d}y \\ &= \int_{-\infty}^{\infty} \delta(x - y) f(y) \, \mathrm{d}y + \frac{i}{\pi} \int_{-\infty}^{\infty} \frac{1}{x - y} f(y) \, \mathrm{d}y \\ &= f(x) + iq(x) \end{aligned}$$

where $q(x)$ is the Hilbert transform of $f(x)$, given by

$$q(x) = \frac{1}{\pi} \int_{-\infty}^{\infty} \frac{f(y)}{x - y} \, \mathrm{d}y$$

Note that the more usual definition of the Hilbert transform is

$$q(x) = \frac{1}{\pi} \int_{-\infty}^{\infty} \frac{f(y)}{y - x} \, \mathrm{d}y$$

However, this definition usually involves more minus signs, and our definition, with $x - y$ in the denominator of the integrand, is generally easier to work with.

2.5.3 The signal attributes

The analytic signal (and thus, the Hilbert transform) is important because it allows us to compute four useful properties of a signal – known as the signal attributes:
Analytic signal

$$s(x) = f(x) + iq(x) \tag{2.16}$$

Amplitude modulations

$$A(x) = \sqrt{f^2(x) + q^2(x)} \tag{2.17}$$

Instantaneous frequency

$$\psi(x) = \int^{x} \frac{1}{A^2} \left(f \frac{\mathrm{d}q}{\mathrm{d}x} - q \frac{\mathrm{d}f}{\mathrm{d}x} \right) \mathrm{d}x \tag{2.18}$$

Unwrapped instantaneous phase

$$\phi(x) = \int^{x} \psi(x) \, \mathrm{d}x \tag{2.19}$$

Frequency modulations

$$\left| \frac{\mathrm{d}}{\mathrm{d}x} \phi(x) \right| \tag{2.20}$$

2.5.4 Computing the analytic signal digitally

$$s(x) = \frac{1}{\pi} \int_{0}^{\infty} F(k) \exp(ikx) \, \mathrm{d}x$$

To compute the Hilbert transform of a discrete function f_n:

1. Take the DFT of f_n to get F_n.
2. Set F_n to zero for all negative frequencies.
3. Compute the inverse DFT of F_n.

The output is as follows: the real part of the DFT is f_n and the imaginary part of the DFT is q_n.

In practice, the DFT can be computed using a FFT. Then the analytic signal $s(x)$ only has frequency components in the positive half of the spectrum. It is therefore referred to as a 'single sideband' signal.

2.5.5 FFT algorithm for computing the Hilbert transform q

Using pseudo-code, the algorithm is:

```
for i=1,2,...,n; do:
sreal(i)=signal(i)
simaginary(i)=0.
 fft(sreal,simaginary)
        for i=1,2,...,n/2; do:
sreal(i)=0.
sreal(i+n/2)=2.*sreal(i+n/2)
simaginary(i)=0.
simaginary(i+n/2)=2.*simaginary(i+n/2)
 inverse_fft(sreal,simaginary)
        for i=1,2,...,n; do:
signal=sreal(i)
hilbert_transform(i)=simaginary(i)
```

2.6 Digital filtering in the frequency domain

Digital filtering in the frequency domain is the basis for an important range of filters. This mode of filtering relies on extensive use of the FFT; without the FFT, it would be impossible to implement. Fourier-based filters usually have a relatively simple algebraic form. They change the 'shape' of the input spectrum via a multiplicative process of the type

$$\text{output spectrum} = \text{filter} \times \text{input spectrum}$$

Filters of this type characterize the *frequency response* of a system to a given input. The algebraic form of the filter usually originates from the solution (with appropriate conditions and approximations) to a particular type of digital signal processing (DSP) problem.

2.6.1 Highpass, lowpass and bandpass filters

An operation that changes the distribution of the Fourier components of a function via a multiplicative process may be defined as a Fourier filtering operation. Thus, in an operation of the form

$$S_i = P_i F_i$$

P_i may be referred to as a filter and S_i can be considered to be a filtered version of F_i.

In general, filters fall into one of the following classes:

- lowpass filters
- highpass filters
- bandpass filters

A *lowpass filter* is one that suppresses or attenuates the high-frequency components of a spectrum while passing the low frequencies within a specified range. A *highpass filter* does exactly the opposite to a lowpass filter – it attenuates the low-frequency components of a spectrum while passing the high frequencies within a specified range. A *bandpass filter* allows only those frequencies within a certain band to pass through. In this sense, lowpass and highpass filters are just special types of bandpass filter. Nearly all signals can be modelled in terms of a bandpass filter that modifies the distribution of the complex Fourier components associated with an information source. Data processing is then required to restore the out-of-band frequencies in order to recover the complex spectrum of the source. This requires a knowledge of the characteristics of the original bandpass filter.

Examples of lowpass digital filters

$$P = \begin{cases} 1, & |f| \leq B \\ 0, & |f| > B \end{cases}$$

The ideal lowpass filter is as shown in Figure 2.3 (the top-hat function). The frequency is f and the bandwidth of the filter is $2B$.

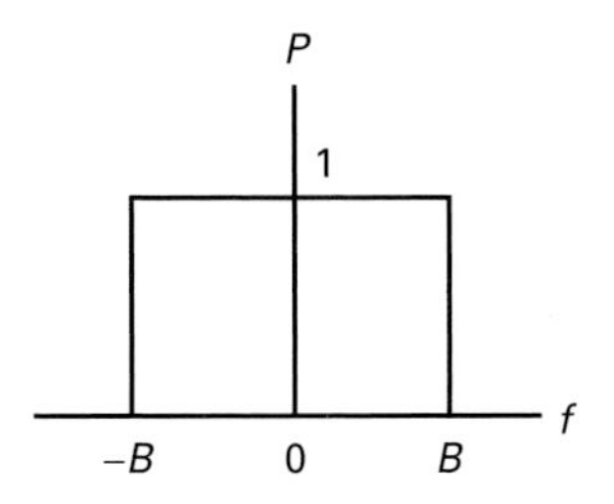

Figure 2.3 Frequency response of an ideal lowpass filter.

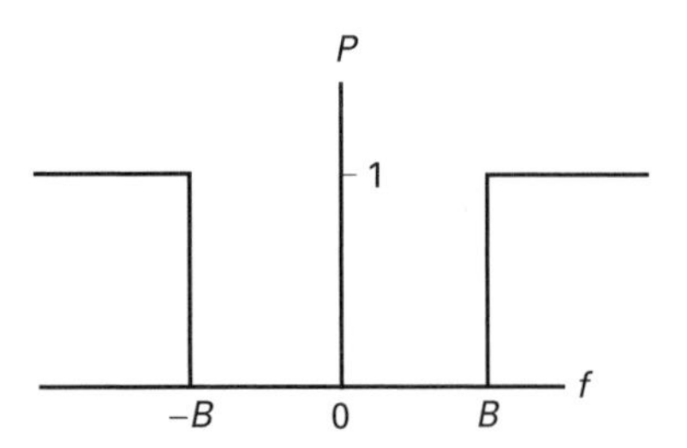

Figure 2.4 Frequency response of an ideal highpass filter.

The Gaussian lowpass filter for sample i is defined as

$$P_i = \exp(-f^2/\sigma^2)$$

where σ is the *standard deviation* of the filter – its halfwidth when $P = \exp(-1)$.

The Butterworth lowpass filter is defined as

$$P_i = \frac{1}{1 + (f/B)^{2n}}, \qquad n = 1, 2, \ldots$$

where B is the cut-off frequency, which defines the bandwidth of the filter and n is the order of the filter, which determines its 'sharpness'.

Examples of highpass digital filters

The ideal highpass filter is defined as (see Figure 2.4)

$$P = \begin{cases} 0, & |f| < B \\ 1, & |f| \geq B \end{cases} \tag{2.21}$$

The Gaussian highpass filter for sample i is defined as

$$P_i = \exp\left(\frac{f^2}{\sigma^2}\right) - 1 \tag{2.22}$$

The Butterworth highpass filter is defined as

$$P_i = \frac{1}{1 + (B/f)^{2n}}, \qquad f \neq 0 \tag{2.23}$$

Examples of bandpass digital filters

The ideal bandpass filter (with $f_c > B$) is defined as (see Figure 2.5)

$$P = \begin{cases} 1, & \text{if } f_c - B \leq f \leq f_c + B \quad \text{or} \quad -f_c - B \leq f \leq -f_c + B \\ 0, & \text{otherwise} \end{cases} \tag{2.24}$$

Here f_c is the centre frequency of the filter, which defines the frequency band, and B defines the bandwidth.

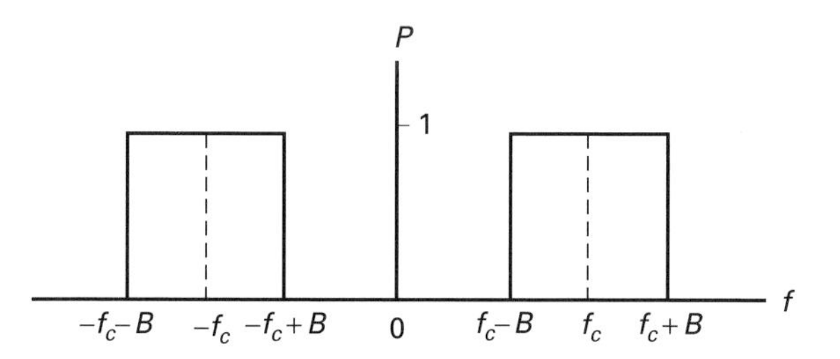

Figure 2.5 Frequency response of ideal bandpass filter.

The quadratic Gaussian bandpass filter is given as

$$P_i = \alpha f^2 \exp\left(\frac{-f^2}{\sigma^2}\right) \tag{2.25}$$

For a given value of α, as σ increases the frequency band over which the filter operates increases.

In general, bandpass filters have at least two control parameters, one to adjust the bandwidth and another to adjust the position of the band.

2.6.2 Inverse filter

The inverse filter is a straightforward approach to deconvolving the equation

$$s_i = p_i \otimes f_i + n_i$$

In the absence of any useful information about the noise n_i, we may ignore it under the assumption that its total contribution to the signal s_i is small. We can then set about inverting the reduced equation

$$s_i = p_i \otimes f_i$$

The basic approach to solving this problem is to process the data s_i in Fourier space. Using the convolution theorem, we have

$$S_i = P_i F_i$$

Reordering and taking the inverse DFT, we get

$$\begin{aligned} f_i &= \text{IDFT}\,(S_i / P_i) \\ &= \text{IDFT}\left(P_i^* S_i / |P_i|^2\right) \end{aligned} \tag{2.26}$$

The function $1/P_i$ is known as the inverse filter.

Criterion for the inverse filter

The criterion for the inverse filter is that the mean square of the noise should be a minimum. In other words, f_i is chosen in such a way that the mean square error

$$e = \|n_i\|^2 = \|s_i - p_i \otimes f_i\|^2$$

is a minimum. Using the orthogonality principle, this error is minimized when

$$(s_i - p_i \otimes f_i) \odot p_i^*(x) = 0$$

Using the correlation and convolution theorems, in Fourier space this equation becomes

$$(S_i - P_i F_i) P_i^* = 0$$

and, solving for F_i, we obtain the same result as before, namely,

$$F_i = \frac{P_i^*}{|P_i|^2} S_i$$

The inverse filter is given by

$$\frac{P_i^*}{|P_i|^2} \tag{2.27}$$

Computational problems

In principle, the inverse filter provides an exact solution to the problem when n_i is close to zero. However, in practice this solution is fraught with difficulties. First, the inverse filter is invariably a singular function owing to the occurrence of zeros in $|P_i|$. Equally bad is the fact that even if the inverse filter is not singular, it is usually ill conditioned. This means that the magnitude of P_i goes to zero so quickly as i increases that $1/|P_i|^2$ rapidly acquires an extremely large value. The effect of this ill conditioning is to amplify the noisy high-frequency components of S_i. This can lead to a reconstruction for f_i that is dominated by the noise in s_i.

The inverse filter can therefore only be used when both the following are true:

- the inverse filter is nonsingular;
- the signal-to-noise ratio of the data is very small.

Such conditions are rare.

The computational problems associated with the inverse filter can be avoided by implementing a variety of different filters whose individual properties and characteristics are suited to certain types of experimental data.

2.6.3 The Wiener filter

The Wiener filter is a minimum mean square filter (i.e. it is based on an application of the least squares principle). This filter is commonly used for signal and image restoration in the presence of additive noise. It is named after the American mathematician Norbert

Wiener who was among the first to discuss its properties. The problem and consequent solution can be formulated using either continuous or discrete functions. Here, the latter approach is taken; this is consistent with the analysis of digital signals.

Statement of the problem

Consider a discrete signal consisting of N real numbers s_i, $i = 0, 1, \ldots, N-1$, each formed by a stationary process of the type (cf. Eqn 2.1a)

$$s_i = \sum_{j=0}^{N-1} p_{i-j} f_j + n_i$$

This can be written as signal = instrument function ⊗ impulse response function + noise

We require an estimate $\hat{f}_i$ for f_i of the form

$$\hat{f}_i = \sum_{j=0}^{N-1} q_j s_{i-j} \tag{2.28}$$

Clearly, the problem is to find q_i. Wiener's solution to this problem is based on utilizing the least squares principle.

2.6.4 Application of the least squares principle

Application of the least squares principle to this class of problems is based on finding a solution for q_i such that

$$e = \| f_i - \hat{f}_i \|^2 \equiv \sum_{i=0}^{N-1} (f_i - \hat{f}_i)^2 \tag{2.29}$$

is a minimum. Under the condition that (for the signal-independent noise)

$$\sum_{i=0}^{N-1} n_{j-i} f_j = 0 \quad \text{and} \quad \sum_{j=0}^{N-1} f_{j-i} n_j = 0$$

the DFT of q_i is given by

$$Q_i = \frac{P_i^*}{|P_i|^2 + |N_i|^2/|F_i|^2} \tag{2.30}$$

where F_i, P_i and N_i are the DFTs of f_i, p_i and n_i respectively. Q_i is known as the Wiener filter and, using the convolution theorem, we can write the required solution as

$$\hat{f}_i = \text{IDFT}\,(Q_i S_i) \tag{2.31}$$

where S_i is the DFT of s_i.

Derivation of the Wiener filter Q_i

Problem Given

$$s_i = \sum_{j=0}^{N-1} p_{i-j} f_j + n_i$$

and

$$\hat{f}_i = \sum_{j=0}^{N-1} q_j s_{i-j}$$

find q_i (and thus Q_i) such that

$$e = \sum_{i=0}^{N-1} (f_i - \hat{f}_i)^2$$

is a minimum.

Solution The functional e defines an error in the sense that the closer $\hat{f}_i$ is to f_i the smaller this error becomes. The error is a functional of the q_i and hence is a minimum when

$$\frac{\partial}{\partial q_k} e(q_k) = 0$$

for all k. Differentiating e, we get

$$\sum_{i=0}^{N-1} \left(f_i - \sum_{j=0}^{N-1} q_j s_{i-j} \right) s_{i-k} = 0$$

We now use the convolution and correlation theorems to write the above equation in the form

$$F_i S_i^* = Q_i S_i S_i^*$$

giving

$$Q_i = \frac{S_i^* F_i}{|S_i|^2}$$

where F_i, S_i and Q_i are the DFTs of f_i, s_i and q_i respectively. This function can be written in terms of P_i and N_i (the DFTs of p_i and n_i respectively), since

$$s_i = \sum_{j=0}^{N-1} p_{i-j} f_j + n_i$$

which transforms to

$$S_i = P_i F_i + N_i$$

via the convolution theorem. This gives

$$S_i^* F_i = (P_i^* F_i^* + N_i^*) F_i$$
$$= P_i^* |F_i|^2 + N_i^* F_i$$
$$|S_i|^2 = S_i S_i^* = (P_i F_i + N_i)(P_i^* F_i^* + N_i^*)$$
$$= |P_i|^2 |F_i|^2 + |N_i|^2 + P_i F_i N_i^* + N_i P_i^* F_i^*$$

and so

$$Q_i = \frac{P_i^* |F_i|^2 + N_i^* F_i}{|P_i|^2 |F_i|^2 + |N_i|^2 + P_i F_i N_i^* + N_i P_i^* F_i^*} \tag{2.32}$$

If we assume that *the noise is signal independent* (a special case) then we can say that, to a good approximation, there is no correlation between the signal and noise and vice versa. This statement is expressed mathematically by the results

$$\sum_{j=0}^{N-1} n_{j-i} f_j = 0 \qquad \text{and} \qquad \sum_{j=0}^{N-1} f_{j-i} n_j = 0$$

Using the correlation theorem, these conditions can be written in the form

$$N_i^* F_i = 0 \qquad \text{and} \qquad F_i^* N_i = 0$$

These conditions allow us to drop the cross terms in the expression for Q_i, leaving us with the result

$$Q_i = \frac{P_i^* |F_i|^2}{|P_i|^2 |F_i|^2 + |N_i|^2}$$

or, after rearranging,

$$Q_i = \frac{P_i^*}{|P_i|^2 + |N_i|^2 / |F_i|^2} \tag{2.33}$$

Properties of the Wiener filter

As the noise goes to zero (i.e. as $|N_i|^2 \to 0$) the Wiener filter, Eqn 2.33, reduces to the inverse filter for the system,

$$\frac{P_i^*}{|P_i|^2}$$

Hence, with minimal noise the Wiener filter behaves like the inverse filter. As the power of the impulse response function goes to zero (i.e. as $|F_i|^2 \to 0$), the Wiener filter has zero gain. This solves any problems concerning the behaviour of the filter as $|P_i|^2$ approaches zero. In other words, the filter is well conditioned.

2.6.5 Practical implementation

As derived in the previous subsection, the Wiener filter is

$$Q_i = \frac{P_i^*}{|P_i|^2 + |N_i|^2/|F_i|^2}$$

The main problem with using it is that, in practice, accurate estimates of $|N_i|^2$ and $|F_i|^2$ are usually not available. So, practical implementation of the Wiener filter normally involves having to make an approximation of the type

$$Q_i \sim \frac{P_i^*}{|P_i|^2 + C}$$

where C is a suitable constant. The value of C ideally reflects knowledge on the SNR of the data, i.e.

$$C \sim \frac{1}{(\text{SNR})^2}$$

In practice, it is not uncommon for a user to apply the Wiener filter over a range of different values of the SNR and then choose a reconstruction $\hat{f}_i$ that is optimal in the sense that it is a good approximation to the user's prior knowledge of the expected form of the impulse response function.

FFT algorithm for the Wiener filter

Clearly, the Wiener filter has a relatively simple algebraic form. The main source of CPU time is the computation of the DFTs. In practice this is done by restricting the data to be of size 2^n and using a FFT. Using pseudo-code, the algorithm for the Wiener filter is:

```
snr=snr*snr
constant=1/snr
 for i=1,2,...,k; do:
sr(i)=signal(i)
si(i)=0.
pr(i)=instfunc(i)
pi(i)=0.
 fft(sr,si)
fft(pr,pi)
 for i=1,2,...,k; do:
denominator=pr(i)*pr(i)+pi(i)*pi(i)+constant
fr(i)=pr(i)*sr(i)+pi(i)*si(i)
fi(i)=pr(i)*si(i)-pi(i)*sr(i)
 fr(i)=fr(i)/denominator
 fi(i)=fi(i)/denominator
 fft(fr,fi)
 for i=1,2,...,k; do:
hatf(i)=fr(i)
```

General statements on the Wiener filter

The Wiener filter is one of the most robust filters for reconstructing signals in the presence of additive noise. It can be used with data of single or dual polarity and for one-dimensional or two-dimensional signal processing problems. A loop can be introduced allowing the user to change the value of the SNR or to sweep through a range of values of the SNR on an interactive basis – 'interactive restoration'.

Techniques for automating filters of this type are of current interest. Alternatively, real-time signal and/or image processing hardware can be used to adjust the SNR value in real time.

2.6.6 Power spectrum equalization

The power spectrum equalization (PSE) filter is based on finding an estimate $\hat{f}_i$ whose power spectrum is equal to the power spectrum of the desired function f_i. The estimate $\hat{f}_i$ is obtained by employing the criterion

$$|F_i|^2 = |\hat{F}_i|^2$$

together with the linear convolution model

$$\hat{f}_i = q_i \otimes s_i$$

Like the Wiener filter, the PSE filter also assumes that the noise is signal independent. Since

$$\hat{F}_i = Q_i S_i = Q_i(P_i F_i + N_i)$$

and, given that $N_i^* F_i = 0$ and $F_i^* N_i = 0$, we have

$$|\hat{F}_i|^2 = \hat{F}_i \hat{F}_i^* = |Q_i|^2(|P_i|^2 |F_i|^2 + |N_i|^2)$$

the PSE criterion can therefore be written as

$$|F_i|^2 = |Q_i|^2(|P_i|^2 |F_i|^2 + |N_i|^2)$$

Solving for $|Q_i|$, $\hat{f}_i$ is then given by

$$\hat{f}_i = \text{IDFT}(Q_i S_i)$$

where Q_i is the PSE filter given by

$$\left(\frac{1}{|P_i|^2 + |N_i|^2/|F_i|^2}\right)^{1/2} \tag{2.34}$$

As for the Wiener filter (Eqn 2.33), in the absence of accurate estimates for $|N_i|^2$ and $|F_i|^2$, we approximate the PSE filter by

$$\left(\frac{1}{|P_i|^2 + C}\right)^{1/2}$$

where

$$C = \frac{1}{\text{SNR}^2}$$

Note that the criterion used to derive this filter can be written in the form

$$\sum_{i=0}^{N-1}(|F_i|^2 - |\hat{F}_i|^2) = 0$$

or, using Parseval's theorem,

$$\sum_{i=0}^{N-1}(|f_i|^2 - |\hat{f}_i|^2) = 0$$

Compare the criterion for the Wiener filter, i.e. that

$$\sum_{i=0}^{N-1}|f_i - \hat{f}_i|^2$$

should be a minimum.

2.6.7 The matched filter

Given that

$$s_i = \sum_{j=0}^{N-1} p_{i-j} f_j + n_i$$

we want to find an estimate $\hat{f}_i$ for the impulse response function:

$$\hat{f}_i = \sum_{j=0}^{N-1} q_j s_{i-j}$$

such that

$$r = \frac{\left|\sum\limits_{i=0}^{N-1} Q_i P_i\right|^2}{\sum\limits_{i=0}^{N-1} |N_i|^2 |Q_i|^2} \tag{2.35}$$

is a maximum. The ratio defining r is a measure of the SNR. In this sense, the matched filter maximizes the SNR of the output.

The matched filter

Assuming that the noise n_i has a 'white' or uniform power spectrum, the filter Q_i that maximizes the SNR defined by r is given by

$$Q_i = P_i^*$$

The required solution is therefore given by

$$\hat{f}_i = \text{IDFT}(P_i^* S_i)$$

Using the correlation theorem we can write

$$\hat{f}_i = \sum_{j=0}^{N-1} p_{j-i} s_j \tag{2.36}$$

The matched filter is based on correlating the signal s_i with the instrument function p_i. The matched filter is frequently used in systems that employ linear frequency-modulated (FM) pulses – 'chirped pulses'.

Derivation of the matched filter

With the problem specified as at the start of this subsection, the matched filter can be seen as essentially a by-product of the *Schwarz inequality*

$$\left|\sum_{i=0}^{N-1} Q_i P_i\right|^2 \leq \sum_{i=0}^{N-1} |Q_i|^2 \sum_{i=0}^{N-1} |P_i|^2 \tag{2.37}$$

The principal trick in deriving the matched filter is to write

$$Q_i P_i = |N_i| Q_i \times \frac{P_i}{|N_i|}$$

so that the above inequality becomes

$$\begin{aligned}\left|\sum_{i=0}^{N-1} Q_i P_i\right|^2 &= \left|\sum_{i=0}^{N-1} |N_i| Q_i \frac{P_i}{|N_i|}\right|^2 \\ &\leq \sum_{i=0}^{N-1} |N_i|^2 |Q_i|^2 \sum_{i=0}^{N-1} \frac{|P_i|^2}{|N_i|^2}\end{aligned}$$

From this result, using the definition of r given in Eqn 2.35, we see that the Schwarz inequality amounts to

$$r \leq \sum_i \frac{|P_i|^2}{|N_i|^2}$$

Now, if r is to be a maximum then we require

$$r = \sum_{i=0}^{N-1} \frac{|P_i|^2}{|N_i|^2}$$

or

$$\left|\sum_{i=0}^{N-1} |N_i| Q_i \frac{P_i}{|N_i|}\right|^2 = \sum_{i=0}^{N-1} |N_i|^2 |Q_i|^2 \sum_{i=0}^{N-1} \frac{|P_i|^2}{|N_i|^2}$$

But this is only true if

$$|N_i|Q_i = \frac{P_i^*}{|N_i|}$$

Hence, r is a maximum when

$$Q_i = \frac{P_i^*}{|N_i|^2} \tag{2.38}$$

If the noise n_i is *white noise* (a special case), then its power spectrum $|N_i|^2$ is uniformly distributed. In particular, under the condition

$$|N_i|^2 = 1 \qquad \forall i = 0, 1, \ldots, N-1$$

then

$$Q_i = P_i^*$$

FFT algorithm for the matched filter

Using pseudo-code, the algorithm for the matched filter is

```
for i=1,2,...,k; do:
sr(i)=signal(i)
si(i)=0.
pr(i)=instfunc(i)
pi(i)=0.
 fft(sr,si)
fft(pr,pi)
 for i=1,2,...,k; do:
fr(i)=pr(i)*sr(i)+pi(i)*si(i)
fi(i)=pr(i)*si(i)-pi(i)*sr(i)
 fft(fr,fi)
 for i=1,2,...,k; do:
        hatf(i)=fr(i)
```

2.6.8 Deconvolution of frequency-modulated signals

The matched filter is frequently used in systems that utilize linear frequency-modulated pulses. As mentioned earlier, pulses of this type are known as chirped pulses. Examples where this particular type of pulse is used include radar, active sonar and some forms of seismic prospecting called vibroseis.

Linear FM pulses

The linear FM pulse is given by

$$p(x) = \exp(i\alpha x^2), \qquad |x| \leq X$$

where α is a constant and X is the length of the pulse. The phase of this pulse is αx^2 and the instantaneous frequency is given by

$$\frac{\mathrm{d}}{\mathrm{d}x}(\alpha x^2) = 2\alpha x$$

which varies linearly with x. Hence, the frequency modulations are linear, which is why the pulse is referred to as a linear FM pulse. In this case, the signal that is recorded is given by (neglecting additive noise)

$$s(x) = \exp(i\alpha x^2) \otimes f(x), \qquad |x| \leq X$$

After matched filtering, we get

$$\hat{f}(x) = \exp(-i\alpha x^2) \odot \exp(i\alpha x^2) \otimes f(x)$$

Note that

$$\exp(-i\alpha x^2) \odot \exp(i\alpha x^2) = \int_{-X/2}^{X/2} \exp[-i\alpha(y+x)^2] \exp(i\alpha y^2)\,\mathrm{d}y$$

$$= \exp(-i\alpha x^2) \int_{-X/2}^{X/2} \exp(-2i\alpha yx)\,\mathrm{d}y$$

Evaluating the integral over y, we have

$$\exp(-i\alpha x^2) \odot \exp(i\alpha x^2) = X \exp(-i\alpha x^2)\,\mathrm{sinc}(\alpha X x)$$

and hence

$$\hat{f}(x) = X \exp(-i\alpha x^2)\,\mathrm{sinc}(\alpha X x) \otimes f(x) \tag{2.39}$$

In some systems the length of the linear FM pulse is relatively long. For example, some radars utilize pulses for which $X \sim 3000$ m; in this case,

$$\cos(\alpha x^2)\,\mathrm{sinc}(\alpha X x) \simeq \mathrm{sinc}(\alpha X x)$$

and

$$\sin(\alpha x^2)\,\mathrm{sinc}(\alpha X x) \simeq 0$$

and so

$$\hat{f}(x) \simeq X\,\mathrm{sinc}(\alpha X x) \otimes f(x)$$

In Fourier space, this last equation can be written as

$$\hat{F}(k) = \begin{cases} (\pi/\alpha) F(k), & |k| \leq \alpha X \\ 0, & \text{otherwise} \end{cases}$$

The estimate $\hat{f}$ is therefore a band-limited estimate of f whose bandwidth is determined by the product of the chirping parameter α and the length of the pulse X.

2.6.9 Constrained deconvolution

Constrained deconvolution provides a filter that gives the user additional control over the deconvolution process. This method is based on minimizing a linear operation on the object f_i of the form $g_i \otimes f_i$, subject to some other constraint. Using the least squares approach, we find an estimate for f_i by minimizing $\|g_i \otimes f_i\|^2$ subject to the constraint

$$\|s_i - p_i \otimes f_i\|^2 = \|n_i\|^2$$

Using this result, we can write

$$\|g_i \otimes f_i\|^2 = \|g_i \otimes f_i\|^2 + \lambda(\|s_i - p_i \otimes f_i\|^2 - \|n_i\|^2)$$

because the quantity inside the parentheses on the right-hand side is zero. The constant λ is called a *Lagrange multiplier.*

Employing the orthogonality principle, $\|g_i \otimes f_i\|^2$ is a minimum when

$$(g_i \otimes f_i) \odot g_i^* - \lambda(s_i - p_i \otimes f_i) \odot p_i^* = 0$$

In Fourier space, this equation becomes

$$|G_i|^2 F_i - \lambda(S_i P_i^* - |P_i|^2 F_i) = 0$$

Solving for F_i, we get

$$F_i = \frac{S_i P_i^*}{|P_i|^2 + \gamma |G_i|^2}$$

where γ is the reciprocal of the Lagrange multiplier ($\gamma = 1/\lambda$). Hence, the constrained least squares filter is given by

$$\frac{P_i^*}{|P_i|^2 + \gamma |G_i|^2} \tag{2.40}$$

The *constrained deconvolution filter* allows the user to change G to suit a particular application. This filter can be thought of as a generalization of the other filters, which can be obtained from (2.40) as follows:

If $\gamma = 0$ then the inverse filter is obtained.
If $\gamma = 1$ and $|G_i|^2 = |N_i|^2/|F_i|^2$ then the Wiener filter is obtained.
If $\gamma = 1$ and $|G_i|^2 = |N_i|^2 - |P_i|^2$ then the matched filter is obtained.

2.6.10 Homomorphic filtering

The homomorphic filter employs the properties of logarithms to write the equation

$$S_i = P_i F_i$$

in the form

$$\ln S_i = \ln P_i + \ln F_i$$

In this case, the object function f_i can be recovered using the formula

$$f_i = \text{IDFT}[\exp(\ln S_i - \ln P_i)]$$

This type of operation is known as homomorphic filtering. In practice, deconvolution by homomorphic processing replaces the problems associated with computing the inverse filter $1/P_i$ by those associated with computing the logarithm of a complex function (i.e. computing the functions $\ln S_i$ and $\ln P_i$). By writing the complex spectra S_i and P_i in terms of their amplitude and phase spectra, we get

$$S_i = A_i^S \exp(i\theta_i^S)$$

and

$$P_i = A_i^P \exp(i\theta_i^P)$$

where A_i^S and A_i^P are the amplitude spectra of S_i and P_i respectively and θ_i^S and θ_i^P are the phase spectra of S_i and P_i respectively. Using these results, we can write

$$f_i = \text{Re IDFT}\left[\exp(\ln A_i^S - \ln A_i^P)\cos(\theta_i^S - \theta_i^P) + i\exp(\ln A_i^S - \ln A_i^P)\sin(\theta_i^S - \theta_i^P)\right] \quad (2.41)$$

2.7 Digital filtering in the time domain

Time domain filtering is based on processing the real-space data of a signal rather than its associated Fourier-space data. There is a wide range of filters of this type but in general they nearly all fall into one of two classes, non-recursive and recursive filters.

2.7.1 Non-recursive filtering

The finite impulse response (FIR) filter is one of the most elementary but widely used filters. An impulse response function is simply the output of the filter when an impulse is applied as input:

$$\begin{array}{ccccc} \text{input} & \longrightarrow & \text{system} & \longrightarrow & \text{output} \\ \delta(x) & \longrightarrow & f(x) & \longrightarrow & s(x) \end{array}$$

We can write

$$s(x) = \int \delta(x-y) f(y)\,\mathrm{d}y = f(x)$$

The general form of an FIR digital filter is

$$s_j = \sum_{i=-N}^{N} p_{j-i} f_i \tag{2.42}$$

Filters of this type have at most $2N + 1$ non-zero coefficients. The discrete and *finite* nature of this filter gives it the name 'finite impulse response'.

2.7.2 FIR filter – discrete convolution

The discrete convolution operation (the convolution sum) can be written in the form

$$s_j = \sum_{i=-N}^{N} p_i f_{j-i} \tag{2.43}$$

where p_i is the digital *impulse response function*.

Consider the case when p_i and f_i are vectors with just three elements:

$$\mathbf{p} = (p_{-1}, p_0, p_1)^T$$
$$\mathbf{f} = (f_{-1}, f_0, f_1)^T, \qquad f_{-2} = f_2 = 0$$

Then, for $j = -1$, Eqn 2.43 gives

$$s_{-1} = \sum_{i=-1}^{1} p_i f_{-1-i} = p_{-1} f_0 + p_0 f_{-1} + p_1 f_{-2} = p_{-1} f_0 + p_0 f_{-1}$$

For $j = 0$,

$$s_0 = \sum_{i=-1}^{1} p_i f_{-i} = p_{-1} f_1 + p_0 f_0 + p_1 f_{-1}$$

For $j = 1$,

$$s_1 = \sum_{i=-1}^{1} p_i f_{1-i} = p_{-1} f_2 + p_0 f_1 + p_1 f_0 = p_0 f_1 + p_1 f_0$$

These results can be written in matrix form as

$$\begin{pmatrix} s_1 \\ s_0 \\ s_1 \end{pmatrix} = \begin{pmatrix} f_0 & f_{-1} & 0 \\ f_1 & f_0 & f_{-1} \\ 0 & f_1 & f_0 \end{pmatrix} \begin{pmatrix} p_{-1} \\ p_0 \\ p_1 \end{pmatrix}$$

Now consider the convolution sum defined as

$$s_j = \sum_{i=-N}^{N} p_{j-i} f_i$$

With

$$\mathbf{p} = (p_{-1}, p_0, p_1)^T, \qquad p_{-2} = p_2 = 0$$
$$\mathbf{f} = (f_{-1}, f_0, f_1)^T$$

we get for $j = -1$,

$$s_{-1} = \sum_{i=-1}^{1} p_{-1-i} f_i = p_0 f_{-1} + p_{-1} f_0 + p_{-2} f_1 = p_0 f_{-1} + p_{-1} f_0$$

For $j = 0$,

$$s_0 = \sum_{i=-1}^{1} p_{-i} f_i = p_1 f_{-1} + p_0 f_0 + p_{-1} f_1$$

For $j = 1$,

$$s_1 = \sum_{i=-1}^{1} p_{1-i} f_i = p_2 f_{-1} + p_1 f_0 + p_0 f_1 = p_1 f_0 + p_0 f_1$$

In matrix form, these results become

$$\begin{pmatrix} s_{-1} \\ s_0 \\ s_1 \end{pmatrix} = \begin{pmatrix} p_0 & p_{-1} & 0 \\ p_1 & p_0 & p_{-1} \\ 0 & p_1 & p_0 \end{pmatrix} \begin{pmatrix} f_{-1} \\ f_0 \\ f_1 \end{pmatrix}$$

Note that

$$\begin{pmatrix} p_0 & p_{-1} & 0 \\ p_1 & p_0 & p_{-1} \\ 0 & p_1 & p_0 \end{pmatrix} \begin{pmatrix} f_{-1} \\ f_0 \\ f_1 \end{pmatrix} = \begin{pmatrix} f_0 & f_{-1} & 0 \\ f_1 & f_0 & f_{-1} \\ 0 & f_1 & f_0 \end{pmatrix} \begin{pmatrix} p_{-1} \\ p_0 \\ p_1 \end{pmatrix}$$

and in general

$$\sum_{i=-N}^{N} p_i f_{j-i} = \sum_{i=-N}^{N} p_{j-i} f_i \tag{2.44}$$

However, the latter definition of a convolution sum is better to work with because it ensures that the matrix elements relate to the impulse response function p_i, i.e.

$$\mathbf{s} = \mathbf{pf}$$

If $\mathbf{f}$ is an Nth-order vector and $\mathbf{p}$ contains just three elements, say, then the convolution sum can be written in the form

$$
\begin{pmatrix} s_{-N} \\ \vdots \\ s_{-1} \\ s_0 \\ s_1 \\ \vdots \\ s_N \end{pmatrix} = \begin{pmatrix} \ddots & & & & & & \\ & \ddots & & & & & \\ & p_1 & p_0 & p_{-1} & & & \\ & & p_1 & p_0 & p_{-1} & & \\ & & & p_1 & p_0 & p_{-1} & \\ & & & & & \ddots & \\ & & & & & & \ddots \end{pmatrix} \begin{pmatrix} f_{-N} \\ \vdots \\ f_{-1} \\ f_0 \\ f_1 \\ \vdots \\ f_N \end{pmatrix} \tag{2.45}
$$

Here, **p** is a tridiagonal matrix. In general the bandwidth of the matrix is determined by the number of elements in the impulse response function.

The inverse process (i.e. given **p** deconvolving **s** to compute **f**) can be solved in this case by using an algorithm for tridiagonal systems of equations.

Useful visualization of the discrete convolution process

Another way of interpreting the discrete convolution process that is useful visually is in terms of two streams of numbers sliding along each other; at each location in the stream, the appropriate numbers are multiplied and the results added together. In terms of the matrix above we have for the right-hand side

$$
\begin{array}{l}
\vdots \\
f_{-4} \\
f_{-3}\, p_1 \\
f_{-2}\, p_0 \quad (= s_{-2}) \\
f_{-1}\, p_{-1} \\
f_0 \\
f_1 \\
f_2\, p_1 \\
f_3\, p_0 \quad (= s_3) \\
f_4\, p_{-1} \\
\vdots
\end{array}
$$

In general, if

$$
\mathbf{f} = \begin{pmatrix} f_{-N} \\ \vdots \\ f_{-1} \\ f_0 \\ f_1 \\ \vdots \\ f_N \end{pmatrix} \qquad \text{and} \qquad \mathbf{p} = \begin{pmatrix} p_{-N} \\ \vdots \\ p_{-1} \\ p_0 \\ p_1 \\ \vdots \\ p_N \end{pmatrix}
$$

then the right-hand side of the generalized Eqn 2.45 is

$$\begin{array}{lll}
\vdots & & \\
f_{-4} & & \\
f_{-3} & \vdots & \\
f_{-2} & p_1 & \\
f_{-1} & p_0 & (= s_{-1}) \\
f_0 & p_{-1} & \\
f_1 & \vdots & \\
f_2 & & \\
f_3 & & \\
f_4 & & \\
\vdots & &
\end{array}$$

Note that the order of the elements of **p** is reversed with respect to those of **f**.

2.7.3 FIR filter – discrete correlation

The discrete correlation operation can be written in the form (cf. Eqn 2.43)

$$s_j = \sum_{i=-N}^{N} p_i f_{i-j} \tag{2.46}$$

where p_i is again the digital impulse response function. Note that compared with the convolution sum the subscript on f is reversed (i.e. f_{j-i} becomes f_{i-j}).

Again consider the case when p_i and f_i are vectors with just three elements:

$$\mathbf{p} = (p_{-1}, p_0, p_1)^T$$
$$\mathbf{f} = (f_{-1}, f_0, f_1)^T, \qquad f_{-2} = f_2 = 0$$

Then, for $j = -1$,

$$s_{-1} = \sum_{i=-1}^{1} p_i f_{i+1} = p_{-1} f_0 + p_0 f_1 + p_1 f_2 = p_{-1} f_0 + p_0 f_1$$

For $j = 0$,

$$s_0 = \sum_{i=-1}^{1} p_i f_i = p_{-1} f_{-1} + p_0 f_0 + p_1 f_1$$

For $j = 1$,

$$s_1 = \sum_{i=-1}^{1} p_i f_{i-1} = p_{-1} f_{-2} + p_0 f_{-1} + p_1 f_0 = p_0 f_{-1} + p_1 f_0$$

As before, this result can be written in matrix form as

$$\begin{pmatrix} s_1 \\ s_0 \\ s_1 \end{pmatrix} = \begin{pmatrix} f_0 & f_1 & 0 \\ f_{-1} & f_0 & f_1 \\ 0 & f_{-1} & f_0 \end{pmatrix} \begin{pmatrix} p_{-1} \\ p_0 \\ p_1 \end{pmatrix}$$

Now consider the correlation sum defined as

$$s_j = \sum_{i-N}^{N} p_{i-j} f_i \tag{2.47}$$

with

$$\mathbf{p} = (p_{-1}, p_0, p_1)^T, \qquad p_{-2} = p_2 = 0$$
$$\mathbf{f} = (f_{-1}, f_0, f_1)^T$$

We get for $j = -1$;

$$s_{-1} = \sum_{i=-1}^{1} p_{i+1} f_i = p_0 f_{-1} + p_1 f_0 + p_2 f_1 = p_0 f_{-1} + p_1 f_0$$

For $j = 0$,

$$s_0 = \sum_{i=-1}^{1} p_i f_i = p_{-1} f_{-1} + p_0 f_0 + p_1 f_1$$

For $j = 1$,

$$s_1 = \sum_{i=-1}^{1} p_{i-1} f_i = p_{-2} f_{-1} + p_{-1} f_0 + p_0 f_1 = p_{-1} f_0 + p_0 f_1$$

In matrix form, this result becomes

$$\begin{pmatrix} s_{-1} \\ s_0 \\ s_1 \end{pmatrix} = \begin{pmatrix} p_0 & p_1 & 0 \\ p_{-1} & p_0 & p_1 \\ 0 & p_{-1} & p_0 \end{pmatrix} \begin{pmatrix} f_{-1} \\ f_0 \\ f_1 \end{pmatrix}$$

Note that for discrete correlation

$$\begin{pmatrix} p_0 & p_1 & 0 \\ p_{-1} & p_0 & p_1 \\ 0 & p_{-1} & p_0 \end{pmatrix} \begin{pmatrix} f_{-1} \\ f_0 \\ f_1 \end{pmatrix} \neq \begin{pmatrix} f_0 & f_1 & 0 \\ f_{-1} & f_0 & f_1 \\ 0 & f_{-1} & f_0 \end{pmatrix} \begin{pmatrix} p_{-1} \\ p_0 \\ p_1 \end{pmatrix}$$

and in general

$$\sum_{i=-N}^{N} p_i f_{i-j} \neq \sum_{i=-N}^{N} p_{i-j} f_i$$

As with the discrete convolution sum, the second definition of a correlation sum, Eqn 2.47, is better to work with because it ensures that the matrix elements relate to the impulse response function p_i, i.e.

$$\mathbf{s} = \mathbf{pf}$$

If $\mathbf{f}$ is an Nth-order vector and $\mathbf{p}$ contains just three elements, say, then the correlation sum can be written in the form

$$\begin{pmatrix} s_{-N} \\ \vdots \\ s_{-1} \\ s_0 \\ s_1 \\ \vdots \\ s_N \end{pmatrix} = \begin{pmatrix} \ddots & & & & & & \\ & \ddots & & & & & \\ & p_{-1} & p_0 & p_1 & & & \\ & & p_{-1} & p_0 & p_1 & & \\ & & & p_{-1} & p_0 & p_1 & \\ & & & & \ddots & & \\ & & & & & & \ddots \end{pmatrix} \begin{pmatrix} f_{-N} \\ \vdots \\ f_{-1} \\ f_0 \\ f_1 \\ \vdots \\ f_N \end{pmatrix} \tag{2.48}$$

Discrete correlation process

As for the discrete convolution process, a useful way of interpreting the discrete correlation process is in terms of two streams of numbers sliding along each other; at each location in the stream, the appropriate numbers are multiplied and the results added together. In terms of the matrix above we have for the right-hand side

$$\begin{array}{l} \vdots \\ f_{-4} \\ f_{-3}\, p_{-1} \\ f_{-2}\, p_0 \quad (= s_{-2}) \\ f_{-1}\, p_1 \\ f_0 \\ f_1 \\ f_2\, p_{-1} \\ f_3\, p_0 \quad (= s_3) \\ f_4\, p_1 \\ \vdots \end{array}$$

In general, if

$$\mathbf{f} = \begin{pmatrix} f_{-N} \\ \vdots \\ f_{-1} \\ f_0 \\ f_1 \\ \vdots \\ f_N \end{pmatrix} \qquad \text{and} \qquad \mathbf{p} = \begin{pmatrix} p_{-N} \\ \vdots \\ p_{-1} \\ p_0 \\ p_1 \\ \vdots \\ p_N \end{pmatrix}$$

then the right-hand side of the generalized Eqn 2.48 is

$$
\begin{array}{lll}
\vdots & & \\
f_{-4} & & \\
f_{-3} & \vdots & \\
f_{-2} & p_{-1} & \\
f_{-1} & p_0 & (= s_{-1}) \\
f_0 & p_1 & \\
f_1 & \vdots & \\
f_2 & & \\
f_3 & & \\
f_4 & & \\
\vdots & &
\end{array}
$$

Unlike in convolution, the order of the elements of **p** is *preserved* with respect to **f**. If the impulse response function is symmetric then the convolution and correlation sums are identical. The terminology associated with the discrete convolution process is also used in the case of discrete correlation. Correlation is also sometimes used in the context of matched filtering.

2.7.4 Computing the FIR

A problem arises in computing the FIR filter (convolution or correlation) at the ends of the array f_i.

Example **p** is a 5×1 kernel:

$$
\begin{array}{ll}
\vdots & \\
f_{N-3} & p_{-2} \\
f_{N-2} & p_{-1} \\
f_{N-1} & p_0 \\
f_N & p_1 \\
 & p_2
\end{array}
$$

In the computation of s_{N-1} there is no number associated with the data f_i with which to multiply p_2. Similarly, in the computation of s_N we have

$$
\begin{array}{ll}
\vdots & \\
f_{N-3} & \\
f_{N-2} & p_{-2} \\
f_{N-1} & p_{-1} \\
f_N & p_0 \\
 & p_1 \\
 & p_2
\end{array}
$$

Here, there are no numbers associated with the array f_i with which to multiply p_1 or p_2. The same situation occurs at the other end of the array f_i. Hence, at both ends of the data, the moving window runs out of data for computing the convolution sum. There are a number of ways of solving this problem, including zero padding, endpoint extension and wrapping. We now describe them.

Zero padding Zero padding assumes that the data is zero beyond the ends of the array, i.e.

$$f_{\pm N\pm 1} = f_{\pm N\pm 2} = f_{\pm N\pm 2} = \cdots = 0$$

Endpoint extension Endpoint extension assumes that the data beyond each end of the array takes on the corresponding end value of the array, i.e. the extrapolated data is equal in value to the endpoints:

$$f_{N+1} = f_{N+2} = f_{N+3} = \cdots = f_N$$

and

$$f_{-N-1} = f_{-N-2} = f_{-N-3} = \cdots = f_{-N}$$

This method is sometimes known as the *constant continuation method.*

Wrapping The wrapping technique assumes that the array is wrapped back on itself so that

$$f_{N+1} = f_{-N}, \qquad f_{N+2} = f_{-N+1}, \qquad f_{N+3} = f_{-N+2}, \qquad \text{etc.}$$

and

$$f_{-N-1} = f_N, \qquad f_{-N-2} = f_{N-1}, \qquad f_{-N-3} = f_{N-2}, \qquad \text{etc.}$$

These methods are used in different circumstances but the endpoint extension technique is probably one of the most widely used.

2.7.5 The moving-average filter

In the moving-average filter, the average value of a set of samples within a predetermined window is computed.

Example Consider a 3 × 1 window:

$$
\begin{array}{l}
\vdots \\
f_i \\
f_{i+1} s_{i+1} = \frac{1}{3}(f_i + f_{i+1} + f_{i+2}) \\
f_{i+2} s_{i+2} = \frac{1}{3}(f_{i+1} + f_{i+2} + f_{i+3}) \\
f_{i+3} s_{i+3} = \frac{1}{3}(f_{i+2} + f_{i+3} + f_{i+4}) \\
f_{i+4} \\
\vdots
\end{array}
$$

As the window moves over the data, the average of the samples 'seen' within the window is computed, hence the term 'moving-average filter'. In mathematical terms, we can express this type of processing in the form

$$s_j = \frac{1}{M} \sum_{j \in w(i)} f_i$$

where $w(i)$ is the window located at i over which the average of the data samples is computed and M is the total number of samples in $w(i)$.

Note that the moving-average filter is just an FIR of the form

$$s_j = \sum_{i=-N}^{N} p_{j-i} f_i$$

with

$$\mathbf{p} = \frac{1}{M}(1, 1, 1, \ldots, 1)$$

So, for a 3 × 1 kernel

$$\mathbf{p} = \frac{1}{3}(1, 1, 1)$$

and for a 5 × 1 kernel

$$\mathbf{p} = \frac{1}{5}(1, 1, 1, 1, 1)$$

This filter can be used to smooth a signal, a feature that can be taken to include the reduction of noise.

2.7.6 The median filter

In the median filter, a window of arbitrary size but usually including an odd number of samples moves over the data, computing the median of the samples defined within each window.

The median The median m of a set of numbers is such that half the numbers in the set are less than m and half are greater than m.

Example 1 If we consider the set (2, 3, 9, 20, 21, 47, 56) then $m = 20$. There are a number of ways to compute the median of an arbitrary set of numbers. One way is to reorder the numbers in ascending values.

Example 2 Reorder (2, 7, 3, 5, 8, 4, 10) by ascending values, obtaining (2, 3, 4, 5, 7, 8, 10), which gives $m = 5$. The reordering of the numbers in this way can be accomplished using a *bubble sort* where the maximum values of the array (in decreasing order) are computed and relocated in a number of successive passes.

Moving-average vs median filter

Both the moving-average and median filters can be used to reduce noise in a signal. Noise reduction algorithms aim to reduce noise while attempting to preserve the information content of a signal. In this sense, because the moving-average filter smoothes the data, the median filter is a superior noise-reducing filter especially in the removal of isolated noise spikes.

REFERENCE

[1] E. C. Ifeachor and B. W. Jervis. *Digital Signal Processing: Practical Approach.* Addison-Wesley, 1993.

3 Statistical analysis, entropy and Bayesian estimation

The processes discussed so far have not taken into account the statistical nature of a speech signal. To do this, another type of approach needs to be considered, based on Bayesian estimation. Bayesian estimation allows digital filters to be constructed whose performance is determined by various parameters that can be determined approximately from the statistics of the data.

3.1 The probability of an event

Suppose we toss a coin, observe whether we get heads or tails and then repeat this process a number of times. As the number of trials increases, we expect that the number of times a head or a tail occurs is half the number of trials. In other words, the probability of getting heads is $1/2$ and the probability of getting tails is also $1/2$. Similarly, if a die with six faces is thrown repeatedly then the probability that it will land on any particular face is $1/6$.

In general, if an experiment is repeated N times and an event x occurs n times, then the probability of this event $P(x)$ is defined as

$$P(x) = \lim_{N \to \infty} \left(\frac{n}{N} \right)$$

The exact probability is the relative frequency of an event as the number of trials tends to infinity. In practice, only a finite number of trials can be conducted and we therefore redefine the probability of an event x as

$$P(x) = \frac{n}{N} \tag{3.1}$$

3.1.1 The joint probability

Suppose we have two coins, which we label A and B. We toss both coins simultaneously N times and record the total number of times that A is heads, the total number of times B is heads and the number of times A and B are heads together. What is the probability that A and B are heads together?

Clearly, if m is the number of times out of N trials that 'heads' occurs simultaneously for A and B then the probability of such an event must be given by

$$P(A \text{ heads and } B \text{ heads}) = \frac{m}{N}$$

This is known as the *joint probability* that A is heads and B is heads.

3.1.2 The conditional probability

Suppose we set up an experiment in which two events A and B can occur. We conduct N trials and record the total number of times A occurs (which is n) and the number of times A and B occur simultaneously (which is m). In this case, the joint probability may be written as

$$P(A \text{ and } B) = \frac{m}{N} = \frac{m}{n} \times \frac{n}{N}$$

Here the quotient n/N is the total probability $P(A)$ that event A occurs. The quotient m/n is the probability that events A and B occur simultaneously given that event A occurs, i.e. that in n events A occurs and in a fraction m/n of these events B occurs also. This latter probability is known as the *conditional probability* and is written as

$$P(B|A) = \frac{m}{n}$$

where the symbol $B|A$ means 'B given A'. Hence, the joint probability can be written as

$$P(A \text{ and } B) = P(A)P(B|A) \tag{3.2}$$

Suppose we do a similar type of experiment but this time we record the total number of times p that event B occurs and the number of times q that event A occurs simultaneously with event B. In this case, the joint probability of events B and A occurring together is given by

$$P(B \text{ and } A) = \frac{q}{N} = \frac{q}{p} \times \frac{p}{N}$$

The quotient p/N is the probability $P(B)$ that event B occurs and the quotient q/p is the probability that events B and A occur simultaneously given that event B occurs. The latter probability is just the probability of getting A 'given' B, i.e.

$$P(A|B) = \frac{q}{p}$$

Therefore, we have

$$P(B \text{ and } A) = P(B)P(A|B) \tag{3.3}$$

3.2 Bayes' rule

The probability that A and B occur together is exactly the same as the probability that B and A occur together, i.e.

$$P(A \text{ and } B) = P(B \text{ and } A)$$

By using the definitions given in Eqns 3.2 and 3.3 of these joint probabilities in terms of the conditional probabilities we arrive at the following formula:

$$P(A)P(B|A) = P(B)P(A|B)$$

or alternatively

$$P(B|A) = \frac{P(B)P(A|B)}{P(A)} \tag{3.4}$$

This result is known as Bayes' rule. It relates the conditional probability of B given A to that of A given B.

Bayesian estimation

In signal analysis, Bayes' rule is written in the form

$$P(f|s) = \frac{P(f)P(s|f)}{P(s)} \tag{3.5}$$

where f is the information we want to recover from a signal sample s, where

$$s = p \otimes f + n$$

and n is the noise. This result is the basis for a class of non-linear restoration methods that are known collectively as Bayes estimators. In simple terms, *Bayesian estimation* attempts to recover f in such a way that the probability of getting f given s is a maximum. In practice this is done by assuming that $P(f)$ and $P(s|f)$ obey certain statistical distributions that are consistent with the experiment in which s is measured. In other words, models are chosen for $P(f)$ and $P(s|f)$ and then f is computed at the point where $P(f|s)$ reaches its maximum value. This occurs when

$$\frac{\partial}{\partial f} P(f|s) = 0$$

The function P is the *probability density function* or PDF; more precisely, the PDF $P(f|s)$ is called the *a posteriori* PDF. Since the logarithm of a function varies monotonically with that function, the *a posteriori* PDF is also a maximum when

$$\frac{\partial}{\partial f} \ln P(f|s) = 0$$

Using Bayes' rule as given in Eqn 3.5, we can write this equation as

$$\frac{\partial}{\partial f} \ln P(s|f) + \frac{\partial}{\partial f} \ln P(f) = 0 \tag{3.6}$$

Because the solution to Eqn 3.6 for f maximizes the *a posteriori* PDF, this method is known as the maximum *a posteriori* (MAP) method.

Example of Bayesian estimation

Suppose we measure a single sample s (one real number) in an experiment where it is known *a priori* that

$$s = f + n$$

n being the noise (a random number). Suppose that it is also known *a priori* that the noise is determined by a Gaussian distribution of the form

$$P(n) = \frac{1}{\sqrt{2\pi\sigma_n^2}} \exp\left(\frac{-n^2}{2\sigma_n^2}\right)$$

where σ_n is the standard deviation of the noise.

The probability of measuring s given f (i.e. the conditional probability $P(s|f)$) is determined by the noise since

$$n = s - f$$

We can therefore write

$$P(s|f) = \frac{1}{\sqrt{2\pi\sigma_n^2}} \exp\left[\frac{-(s-f)^2}{2\sigma_n^2}\right]$$

To find the MAP estimate, the PDF for f is needed. Suppose that f also has a zero-mean Gaussian distribution, of the form

$$P(f) = \frac{1}{\sqrt{2\pi\sigma_f^2}} \exp\left(\frac{-f^2}{2\sigma_f^2}\right)$$

Then use Eqn 3.6 to obtain

$$\frac{\partial}{\partial f} \ln P(s|f) + \frac{\partial}{\partial f} \ln P(f) = \frac{(s-f)}{\sigma_n^2} - \frac{f}{\sigma_f^2} = 0$$

Solving this equation for f gives

$$f = \frac{s\tau^2}{1+\tau^2}$$

where τ is the signal-to-noise ratio (SNR):

$$\tau = \frac{\sigma_f}{\sigma_n}$$

Note that:

1. As $\sigma_n \to 0$, $f \to s$, as required since $s = f + n$ and n has a zero-mean Gaussian distribution.
2. The solution we acquire for f is entirely dependent on prior information about the PDF for f. A different PDF produces an entirely different solution.

Now suppose that f obeys a Rayleigh distribution of the form

$$P(f) = \begin{cases} \dfrac{f}{\sigma_f^2} \exp\left(\dfrac{-f^2}{2\sigma_f^2}\right), & f \geq 0 \\ 0, & f < 0 \end{cases}$$

In this case,

$$\frac{\partial}{\partial f} \ln P(f) = \frac{1}{f} - \frac{f}{\sigma_f^2}$$

and we get (still assuming that the noise obeys the same zero-mean Gaussian distribution)

$$\frac{s - f}{\sigma_n^2} + \frac{1}{f} - \frac{f}{\sigma_f^2} = 0$$

This equation is quadratic in f. Solving it, we obtain

$$f = \frac{s\tau^2}{2(1 + \tau^2)} \left[1 \pm \sqrt{1 + \frac{4\sigma_n^2}{s^2\tau^2}\left(1 + \frac{1}{\tau^2}\right)}\right]$$

The solution for f that maximizes the value of $P(f|s)$ can then be written in the form

$$f = \frac{s}{2a}\left(1 + \sqrt{1 + \frac{4a\sigma_n^2}{s^2}}\right) \tag{3.7}$$

where

$$a = 1 + \frac{1}{\tau^2}$$

This is a non-linear estimate for f. If

$$\frac{2\sigma_n\sqrt{a}}{s} \ll 1$$

then

$$f \simeq \frac{s}{a} \tag{3.8}$$

In this case, f is linearly related to s. Note that this linearized estimate is identical to the MAP estimate obtained earlier, where it was assumed that f had a Gaussian distribution.

3.3 Maximum-likelihood estimation

From the above example, it is clear that Bayesian estimation (i.e. the MAP method) is only as good as the prior information on the statistical behaviour of f – the object for which we seek a solution. When $P(f)$ is broadly distributed compared with $P(s\,|\,f)$, the peak value of the *a posteriori* PDF will lie close to the peak value of $P(f)$. In particular, if $P(f)$ is roughly constant then $\partial \ln P(f)/\partial f$ is close to zero and therefore, from Eqn 3.6,

$$\frac{\partial}{\partial f} \ln P(f\,|s) \simeq \frac{\partial}{\partial f} \ln P(s\,|\,f)$$

In this case, the *a posteriori* PDF is a maximum when

$$\frac{\partial}{\partial f} \ln P(s\,|\,f) = 0 \tag{3.9}$$

The estimate for f that is obtained by solving this equation for f is called the maximum-likelihood or ML estimate. To obtain this estimate, prior knowledge of only the statistical fluctuations of the conditional probability is required. If, as in the previous example, we assume that the noise is a zero-mean Gaussian distribution then the ML estimate is given by

$$f = s \tag{3.10}$$

Note that this is the same as the MAP estimate when the standard deviation of the noise is zero.

3.3.1 ML verses MAP estimation

The basic difference between the MAP and ML estimates is that the ML estimate ignores prior information about the statistical fluctuations of the object f. Maximum-likelihood estimation requires only a model for the statistical fluctuations of the noise. For this reason, the ML estimate is usually easier to compute. It is also the estimate to use in cases where there is a complete lack of knowledge about the statistical behaviour of the object.

A further example of Bayesian estimation

To illustrate further the difference between the MAP and ML estimates and to show their use in signal analysis, consider the case where we measure N samples of a real signal s in the presence of additive noise n; s is the result of transmitting a known signal f modified by a random amplitude factor a. The samples of the signal are given by

$$s_i = af_i + n_i, \qquad i = 1, 2, \ldots, N \tag{3.11}$$

The problem is to find an estimate for a. To solve problems of this type using Bayes estimation, we must introduce multidimensional probability theory. In this case, the PDF is a function of not just one number s but a set of numbers $s_1, s_2, \ldots, s_N$. It is therefore a function in a vector space. To emphasize this, we use the vector notation

$$P(\mathbf{s}) \equiv P(s_1, s_2, \ldots, s_N)$$

The *ML estimate* is found by solving the equation (see Eqn 3.9)

$$\frac{\partial}{\partial a} \ln P(\mathbf{s}|a) = 0 \tag{3.12}$$

for a, where

$$P(\mathbf{s}|a) \equiv P(s_1, s_2, \ldots, s_N|a)$$

Assume now that the noise is described by a zero-mean Gaussian distribution of the form

$$P(\mathbf{n}) \equiv P(n_1, n_2, \ldots, n_N) = \frac{1}{\sqrt{2\pi\sigma_n^2}} \exp\left(-\frac{1}{2\sigma_n^2}\sum_{i=1}^{N} n_i^2\right)$$

Using Eqn 3.11, the conditional probability is then given by

$$P(\mathbf{s}|a) = \frac{1}{\sqrt{2\pi\sigma_n^2}} \exp\left(-\frac{1}{2\sigma_n^2}\sum_{i=1}^{N} (s_i - af_i)^2\right)$$

and so, from Eqn 3.12,

$$\frac{\partial}{\partial a} \ln P(\mathbf{s}|a) = \frac{1}{\sigma_n^2}\sum_{i=1}^{N} (s_i - af_i) f_i = 0$$

Solving this last equation for a we obtain the ML estimate

$$a = \frac{\sum\limits_{i=1}^{N} s_i f_i}{\sum\limits_{i=1}^{N} f_i^2} \tag{3.13}$$

The *MAP estimate* is obtained by solving the equation

$$\frac{\partial}{\partial a} \ln P(\mathbf{s}|a) + \frac{\partial}{\partial a} \ln P(a) = 0$$

for a. Using the same distribution for the conditional PDF, let us assume that a has a zero-mean Gaussian distribution of the form

$$P(a) = \frac{1}{\sqrt{2\pi\sigma_a^2}} \exp\left(\frac{-a^2}{2\sigma_a^2}\right)$$

where σ_a^2 is the standard deviation of a. In this case,

$$\frac{\partial}{\partial a}\ln P(a) = -\frac{a}{\sigma_a^2}$$

and, hence, the MAP estimate is obtained by solving the equation

$$\frac{\partial}{\partial a}\ln P(\mathbf{s}|a) + \frac{\partial}{\partial a}\ln P(a) = \frac{1}{\sigma_n^2}\sum_{i=1}^{N}(s_i - af_i)f_i - \frac{a}{\sigma_a^2} = 0$$

for a. The solution to this equation is given by

$$a = \frac{\left(\sigma_a^2/\sigma_n^2\right)\sum\limits_{i=1}^{N} s_i f_i}{1 + \left(\sigma_a^2/\sigma_n^2\right)\sum\limits_{i=1}^{N} f_i^2} \tag{3.14}$$

Note, that if $\sigma_a \gg \sigma_n$ then

$$a \simeq \frac{\sum\limits_{i=1}^{N} s_i f_i}{\sum\limits_{i=1}^{N} f_i^2} \tag{3.15}$$

which is the same as the ML estimate, Eqn 3.13.

3.4 The maximum-likelihood method

The maximum-likelihood method uses the principles of Bayesian estimation, discussed in the two previous subsections, to design deconvolution algorithms. The problem is as follows (note that all discrete functions are assumed real). Given the digital signal

$$s_i = \sum_{j=0}^{N-1} p_{i-j} f_j + n_i \tag{3.16}$$

find an estimate for f_i when p_i and the statistics for n_i are known.

The ML estimate for f_i is determined by solving the equation

$$\frac{\partial}{\partial f_k}\ln P(s_1, s_2, \ldots, s_N | f_1, f_2, \ldots, f_N) = 0 \tag{3.17}$$

As before, the algebraic form of the estimate depends upon the model that is chosen for the PDF. We assume again that the noise has a zero-mean Gaussian distribution. In this case, the conditional PDF is given by

$$P(\mathbf{s}|\mathbf{f}) = \frac{1}{\sqrt{2\pi\sigma_n^2}}\exp\left[-\frac{1}{2\sigma_n^2}\sum_{i=0}^{N-1}\left(s_i - \sum_{j=0}^{N-1} p_{i-j} f_j\right)^2\right]$$

where σ_n is the standard deviation of the noise. Substituting this result into Eqn 3.17 we get

$$\frac{1}{\sigma_n^2}\sum_{j=0}^{N-1}(s_i - p_{i-j}f_j)p_{i-k} = 0$$

or

$$\sum_i s_i p_{i-k} = \sum_i \left(\sum_j p_{i-j}f_j\right) p_{i-k}$$

Using appropriate symbols, we may write this equation in the form

$$s_k \odot p_k = (p_k \otimes f_k) \odot p_k$$

where $\odot$ and $\otimes$ denote the correlation and convolution sums respectively.

The ML estimate is now obtained by solving the above equation for f_k. This can be done by transforming it into Fourier space. Using the correlation and convolution theorems, in Fourier space this equation becomes

$$S_k P_k = (P_k F_k)P_k$$

and thus

$$f_k = \text{IDFT}(F_k) = \text{IDFT}\left(\frac{S_k P_k^*}{|P_k|^2}\right) \tag{3.18}$$

For Gaussian statistics, the ML filter is given by

$$\frac{P_k^*}{|P_k|^2} \tag{3.19}$$

which is identical to the inverse filter.

3.5 The maximum *a posteriori* method

The maximum *a posteriori* method is based on computing the f_i such that

$$\frac{\partial}{\partial f_k}\ln P(s_1, s_2, \ldots, s_N | f_1, f_2, \ldots, f_N) + \frac{\partial}{\partial f_k}\ln P(f_1, f_2, \ldots, f_N) = 0 \tag{3.20}$$

for all k. Consider the following models for the relevant PDF.

1. Gaussian statistics for the noise:

$$P(\mathbf{s}|\mathbf{f}) = \frac{1}{\sqrt{2\pi\sigma_n^2}}\exp\left(\frac{1}{2\sigma_n^2}\sum_i \left|s_i - \sum_j p_{i-j}f_j\right|^2\right)$$

2. Zero-mean Gaussian distribution for the object:

$$P(\mathbf{f}) = \frac{1}{\sqrt{2\pi\sigma_f^2}} \exp\left(-\frac{1}{2\sigma_f^2}\sum_i |f_i|^2\right)$$

By substituting these expressions for $P(\mathbf{s}|\mathbf{f})$ and $P(\mathbf{f})$ into Eqn 3.20, we obtain

$$\frac{1}{\sigma_n^2}\sum_i \left(s_i - \sum_j p_{i-j} f_j\right) p_{i-k}^* - \frac{1}{\sigma_f^2} f_k = 0$$

Rearranging, we may write this result in the form

$$s_k \odot p_k^* = \frac{\sigma_k^2}{\sigma_f^2} f_k + (p_k \otimes f_k) \odot p_k^*$$

In Fourier space, this equation becomes

$$S_k P_k^* = \frac{1}{\tau^2} F_k + |P_k|^2 F_k$$

The MAP filter for Gaussian statistics is therefore given by

$$\text{MAP filter} = \frac{P_k^*}{|P_k|^2 + 1/\tau^2} \tag{3.21}$$

where

$\tau = \sigma_f / \sigma_n$ is the SNR.

Note that this filter is the same as the Wiener filter under the assumption that the power spectra of the noise and object are constant. Also, note that

$$\lim_{\sigma_n \to 0} (\text{MAP filter}) = \text{ML filter} \tag{3.22}$$

3.6 The maximum-entropy method

The entropy of a system describes its disorder – it is a measure of the lack of knowledge about the exact state of a system. In signal analysis, the entropy of a signal is a measure of the lack of knowledge about the information content of the signal. Noisy signals will in general have a large entropy. In order to recover a signal in the presence of noise, we can design an algorithm based on an estimate having minimum entropy. The general definition for the entropy of a system E is

$$E = -\sum_i P_i \ln P_i$$

where P_i is the probability that the system is in a state i.

In maximum-entropy deconvolution, a reconstruction for f_i is found such that

$$E = \sum_i f_i \ln f_i$$

is a minimum or, equivalently, such that

$$E = -\sum_i f_i \ln f_i$$

is a maximum. Note that because of the nature of the logarithmic function this type of solution must be restricted to cases where f_i is real and positive.

The signal s_i is given by

$$s_i = \sum_j p_{i-j} f_j + n_i$$

We can therefore write the entropy as

$$E = \sum_i f_i \ln f_i + \lambda \left(\sum_i \left| s_i - \sum_j p_{i-j} f_j \right|^2 - \sum_i |n_i|^2 \right)$$

where λ is a Lagrange multiplier. This expression for the entropy is a maximum when

$$\frac{\partial E}{\partial f_i} = 0$$

for all i. Differentiating, we obtain

$$1 + \ln f_i - \lambda (s_i \odot p_i^* - p_i \otimes f_i \odot p_i^*) = 0$$

or, after rearranging,

$$f_i = \exp[-1 + \lambda (s_i \odot p_i^* - p_i \otimes f_i \odot p_i^*)] \tag{3.23}$$

The last equation is *transcendental* in f_i and, as such, requires that f_i is evaluated iteratively, i.e. using

$$f_i^{k+1} = \exp[-1 + \lambda (s_i \odot p_i^* - p_i \otimes f_i^k \odot p_i^*)]$$

The rate of convergence of this solution is determined by the value of the Lagrange multiplier. The efficiency and overall success of maximum-entropy deconvolution therefore depends on the value of the Lagrange multiplier used.

The iterative nature of this solution is undesirable because it is time consuming and may require many iterations before convergence is achieved. For this reason a linear approximation is useful. This is obtained by retaining just the first two terms (i.e. the linear terms) in the Taylor series expansion of the exponential function (Eqn 3.23), which gives the following equation:

$$f_i = \lambda(s_i \odot p_i^* - p_i \otimes f_i \odot p_i^*)$$

In Fourier space, this equation becomes

$$F_i = \lambda S_i P_i^* - \lambda |P_i|^2 F_i$$

Solving for F_i, we get

$$F_i = \frac{S_i P_i^*}{|P_i|^2 + 1/\lambda}$$

The linearized maximum-entropy filter is therefore given by

$$\frac{P_i^*}{|P_i|^2 + 1/\lambda} \tag{3.24}$$

Note that the algebraic form of this function is very similar to the MAP filter, in the sense that $1/|P_i|^2$ is regularized by a defined constant.

3.7 Spectral extrapolation

The effect of deconvolving a signal is to recover the information it contains by compensating for the blur caused by the instrument function. The resolution of the information obtained by this process is determined by the bandwidth of the data, which, in turn, is controlled by the finite frequency response of the instrument function.

3.7.1 Band-limited functions

A band-limited function is a function whose spectral bandwidth is finite; most real signals are band-limited functions. The bandwidth determines the resolution of a signal. This leads one to consider the problem of how the bandwidth, and hence the resolution of the signal, can be increased synthetically. In other words, how can we extrapolate the spectrum of a band-limited function from an incomplete sample? Solutions to this type of problem are important in signal analysis when a resolution is required which is not a characteristic of the signal provided and which is difficult or even impossible to achieve experimentally. The type of resolution obtained by extrapolating the spectrum of a band-limited function is referred to as super-resolution, and the process is known as *spectral extrapolation*.

3.8 Formulation of the problem

The basic problem is an inverse problem. In its simplest form, it is concerned with the inversion of the integral equation

$$s(x) = \int_{-\infty}^{\infty} f(y) \frac{\sin[K(x-y)]}{\pi(x-y)} \, \mathrm{d}y \tag{3.25}$$

for f, where K determines the bandwidth of the signal s and hence the resolution of f. This equation is a convolution over the interval $[-\infty, \infty]$. Hence, we may view our problem (i.e. the super-resolution problem) in terms of deconvolving s to recover the object f in the special case when the instrument function is a sinc function.

Eigenfunction solution

In practice, signals have a finite duration and so

$$f(x) = 0, \qquad |x| > X$$

In this case, we can restrict the convolution integral to the interval $[-X, X]$ and model the signal as

$$s(x) = \int_{-X}^{X} f(y) \frac{\sin[K(x-y)]}{\pi(x-y)} \, \mathrm{d}y$$

The object can be expressed in the following form:

$$f(x) = \sum_{n=0}^{\infty} \lambda_n^{-1} \left[\int_{-X}^{X} s(y)\phi_n(y) \, \mathrm{d}y \right] \phi_n(x)$$

where the eigenfunctions ϕ_n are the prolate spheroidal wave functions given by the solution to the equation

$$\int_{-X}^{X} \phi_n(y) \frac{\sin[K(x-y)]}{\pi(x-y)} \, \mathrm{d}y = \lambda_n \phi_n(x)$$

and λ_n are the associated eigenvalues. Like other theoretical inverse solutions, this solution is extremely sensitive to errors in measuring s (i.e. experimental noise). It is therefore often difficult to achieve a stable solution using this method with real signals.

3.8.1 Solution by analytic continuation

Using the convolution theorem, we can write Eqn 3.25 in Fourier space as

$$S(k) = H(k)F(k), \qquad |k| \leq \infty$$

where

$$H(k) = \begin{cases} 1, & |k| \leq K \\ 0, & |k| > K \end{cases}$$

or alternatively

$$S(k) = \begin{cases} F(k), & |k| \leq K \\ 0, & \text{otherwise} \end{cases}$$

Here, S and F are the Fourier transforms of s and f respectively. In this form, our problem is to recover $F(k)$ for all values of k from $S(k)$. Because f has finite support, its spectrum is analytic and therefore can be analytically continued, in principle, beyond $[-K, K]$ to provide higher resolution.

This can be done by computing the Taylor series for F, i.e.

$$F(k) = \sum_{n=-\infty}^{\infty} F^{(n)}(0)\frac{k^n}{n!}$$

The derivatives $F^{(n)}$ of F at $k = 0$ can be determined from the finite segment $F(k)$, $|k| \leq K$, which is equal to S. Hence, we can write

$$F(k) = \sum_{n=-\infty}^{\infty} S^{(n)}(0)\frac{k^n}{n!} \tag{3.26}$$

This method of extrapolation is known as analytic continuation. Once again, although of theoretical interest, in practice this method is fraught with problems. First, it is not possible to evaluate $S^{(n)}(0)$ accurately when the signal is noisy. Second, the truncation of the Taylor series (which is necessary in practice) yields large errors for large k and, since knowledge of $F(k)$ is required for all values of k, errors of this kind are unacceptable. Thus, analytic continuation fails in the presence of even small amounts of noise.

3.9 Reformulation of the problem

There are two important features of the equation

$$s(x) = \int_{-\infty}^{\infty} f(y)\frac{\sin[K(x-y)]}{\pi(x-y)}\,dy$$

and therefore of its inversion, which in practice are entirely unsuitable:

1. it is assumed that the signal s can be measured without any experimental error;
2. it is assumed that all the functions are continuous.

In fact, we are usually dealing with a digital signal that is a discrete set of real or complex numbers. From a digital signal, we can generate discrete Fourier data (via the discrete Fourier transform). These data are related to s via the transform

$$S_n \equiv S(k_n) = \int_{-X}^{X} s(x)\exp(-ik_n x)\,\mathrm{d}x, \quad |k_n| \le K \tag{3.27}$$

where

$$s(x) = f(x) + n(x)$$

The data is a set of N numbers and define the band-limited signal

$$s_{BL}(x) = \sum_{n=-N/2}^{N/2} S_n \exp(ik_n x) \tag{3.28}$$

This signal may be complex or real and of alternating or fixed polarity, depending on the type of experiment that is being conducted. In each case, the problem is to reconstruct the object f from N spectral samples in the presence of additive noise n.

3.10 Ill-posed problems

There is no exact, unique or even correct solution to the type of problem that has been presented here. In other words, it is simply not possible to derive such a solution for f from S_n (Eqn 3.27); it is only possible to derive an estimate for it. There are two reasons for this.

1. The exact value of the noise n at x is not known, only (at best) the probability that n has a particular value at x.
2. Even in the absence of noise, this type of problem is ill posed, as we now discuss.

A problem is well posed if the solution

- exists
- is unique
- depends continuously on the data

If a problem violates any of these conditions, then it is ill posed. It is the third condition that causes the main problem with digital signals. The finite nature of the data means that there are many permissible solutions to the problem. As a consequence, we are faced with the problem of having to select one particular reconstruction. To overcome this inherent ambiguity, prior knowledge must be used to reduce the class of allowed solutions. For this reason, the use of prior information in the treatment of ill-posed problems of this nature is essential. In addition to prior information, the discrete nature of the data forces one to employ mathematical models for f. In principle, an unlimited number of different models can be used, which accounts for the wide variety and diversity of algorithms that have been designed to cope with problems of this kind. Since all such algorithms attempt to solve the same basic problem, attention should focus on designs which are simple to implement and compute and which are data adaptive and reliable in the presence of noise of a varying dynamic range.

Because sampled data are always insufficient to specify a unique, correct solution and because no algorithm is able to reconstruct equally well all characteristics of the object, it is essential that the user is able to play a role in the design of the algorithm that is employed and to incorporate maximum knowledge of the expected features in the object. This allows optimum use to be made of the available data together with the user's experience, judgement and intuition. Models for the object, and conditions for the reconstruction, that are amenable to modification as knowledge about the object improves should be utilized. This provides the opportunity for the user to participate in the design of an algorithm by choosing the conditions for the reconstruction that are best suited to a particular application.

3.11 The linear least squares method

Given Eqn 3.27,

$$S_n \equiv S(k_n) = \int_{-X}^{X} s(x) \exp(-ik_n x)\, dx, \qquad |k_n| \leq K$$

for some data S_n, the problem is to solve for f. We start by considering a linear polynomial model for $f(x)$ of the following form,

$$f(x) = \sum_n A_n \exp(ik_n x) \tag{3.29}$$

where

$$\sum_n \equiv \sum_{n=-N/2}^{N/2}$$

This model for f is just a Fourier series. In order to compute f, the coefficients A_n (which are complex numbers) must be known. Given the model above, our problem is reduced to finding a method of computing the A_n. How this is done depends on the criterion for the reconstruction that is chosen. The choice depends on a number of factors, such as the nature of the data, the complexity of the resulting algorithm and its computational cost. Here we consider a least squares approach. The application of this approach for spectral extrapolation is known as the Gerchberg–Papoulis method.

3.12 The Gerchberg–Papoulis method

In this case, the A_n are chosen in such a way that the mean square error

$$e = \int_{-X}^{X} |s(x) - f(x)|^2 \mathrm{d}x \tag{3.30}$$

is a minimum. Since $s = f + n$ this is equivalent to minimizing the noise in the signal. Substituting the equation

$$f(x) = \sum_n A_n \exp(ik_n x)$$

into the above equation and differentiating with respect to A_m we get (using the orthogonality principle)

$$\frac{\partial e}{\partial A_m} = -\int_{-X}^{X} \left[s - \sum_n A_n \exp(ik_n x)\right] \exp(-ik_m x)\,\mathrm{d}x = 0$$

Note that the above result is obtained using

$$\frac{\partial e}{\partial \operatorname{Re} A_m} = 0, \qquad \frac{\partial e}{\partial \operatorname{Im} A_m} = 0$$

Interchanging the order of integration and summation, we obtain the following equation:

$$\int_{-X}^{X} s(x) \exp(-ik_m x)\,\mathrm{d}x = \sum_n A_n \int_{-X}^{X} \exp[-i(k_m - k_n)x]\,\mathrm{d}x \tag{3.31}$$

The left-hand side of this equation is just the discrete Fourier data that is provided, $S(k_n)$, and the integral on the right-hand side gives a sinc function,

$$\int_{-X}^{X} \exp[-i(k_m - k_n)x]\,\mathrm{d}x = 2X \operatorname{sinc}[(k_m - k_n)X]$$

By solving Eqn 3.31 for A_n, the object function $f(x)$ can be obtained.

3.12.1 Weighting functions

This solution of Eqn 3.31 gives a least squares approximation for $f(x)$. To compute A_n from $S(k_n)$, the value of X, the support of the object needs to be known. We can write $f(x)$ in the closed form

$$f(x) = w(x) \sum_n A_n \exp(ik_n x)$$

if we introduce the weighting function

$$w(x) = \begin{cases} 1, & |x| \le X \\ 0, & |x| > X \end{cases}$$

This function is a simple form of *a priori* information. In this case it is information about the finite extent of the object.

3.12.2 Incorporation of *a priori* information

The algebraic form of the equation

$$f(x) = w(x) \sum_n A_n \exp(ik_n x)$$

suggests that the function $w(x)$ can be used to incorporate more general prior knowledge about the object.

Consider the case where we are given data $S(k_n)$, defined by the equation

$$S_n \equiv S(k_n) = \int_{-X}^{X} s(x) \exp(-ik_n x)\, \mathrm{d}x, \qquad |k_n| \leq K$$

together with some form of prior knowledge on the structure of $f(x)$ that can be used to construct a suitable weighting function $w(x)$. The weighting function can be used to compute $f(x)$ as follows:

$$f(x) = w(x) \sum_n A_n \exp(ik_n x)$$

Substituting these equations into

$$e = \int_{-X}^{X} |s(x) - f(x)|^2 \mathrm{d}x$$

we find that the error e is a minimum when

$$\int_{-X}^{X} s(x) w(x) \exp(-ik_m x)\, \mathrm{d}x = \sum_n A_n \int_{-X}^{X} [w(x)]^2 \exp[-i(k_m - k_n)x]\, \mathrm{d}x \tag{3.32}$$

A problem occurs: for arbitrary functions w, which is what we must assume if different types of prior information are to be incorporated, the integral on the left-hand side of the above equation is not the same as the data-provided $S(k_n)$. In other words, Eqn 3.32 cannot be solved from the available data; it is not 'data-consistent'.

A way of overcoming this difficulty is to modify the expression for the mean square error function and introduce the following 'inverse weighted' form:

$$e = \int_{-X}^{X} |s(x) - f(x)|^2 \frac{1}{w(x)}\, \mathrm{d}x$$

The error e lies in a weighted Hilbert space designed to provide data consistency. It is a minimum when

$$\int_{-X}^{X} s(x) \exp(-ik_m x)\, \mathrm{d}x = \sum_n A_n \int_{-X}^{X} w(x) \exp[-i(k_m - k_n)x]\, \mathrm{d}x$$

Here, the data on the left-hand side of the equation is equal to $S(k_m)$. We therefore have a data-consistent equation of the form

$$S(k_m) = \sum_n A_n W(k_m - k_n) \tag{3.33}$$

where

$$W(k_m) = \int_{-X}^{X} w(x) \exp(-ik_m x)\,\mathrm{d}x, \qquad |k_m| \leq K$$

This method provides a solution that allows arbitrary weighting functions $w(x)$ containing additional prior information on the structure of f to be introduced. The method can be summarized as follows.

1. Given the data $S(k_n)$, construct a weighting function $w(x)$ that is obtained from prior knowledge on the structure of $f(x)$.
2. Compute $W(k_n)$ from $w(x)$.
3. Solve the equation

   $$\sum_n A_n W(k_m - k_n) = S(k_m)$$

   to obtain the coefficients A_n.
4. Compute the estimate $w(x) \sum_n A_n \exp(ik_n x)$.

This algorithm is based on minimizing the inverse weighted mean square error function given by

$$e = \int_{-X}^{X} |s(x) - f(x)|^2 \frac{1}{w(x)}\,\mathrm{d}x \tag{3.34}$$

Note that the algebraic form of this error indicates that w should be greater than zero to avoid singularities in $1/w$.

3.12.3 Practical considerations

In practice, the data $S(k_n)$ is obtained by taking the discrete Fourier transform of some band-limited signal $s_{BL}(x)$. The data $W(k_n)$ is obtained by computing the discrete Fourier transform of $w(x)$ and reducing the bandwidth of the spectrum so that it is the same as that for $S(k_n)$. We then solve

$$\sum_n A_n W(k_m - k_n) = S(k_m)$$

for A_n. This equation is just a discrete convolution in Fourier space, i.e.

$$S(k_n) = A(k_n) \otimes W(k_n)$$

where $\otimes$ denotes the convolution sum. Hence, using the convolution theorem, we can write

$$s_{BL}(x) = a(x)w_{BL}(x)$$

where s_{BL} and w_{BL} are band-limited estimates of $s(x)$ and $w(x)$ respectively, given by

$$s_{BL}(x) = \sum_n S(k_n)\exp(ik_n x)$$

and

$$w_{BL}(x) = \sum_n W(k_n)\exp(ik_n x)$$

and where

$$a(x) = \sum_n A(k_n)\exp(ik_n x)$$

Using the (weighted) model for $f(x)$ we have

$$\begin{aligned} f(x) &= w(x)\sum_n A_n \exp(ik_n x) \\ &= w(x)\sum_n A(k_n)\exp(ik_n x) = w(x)a(x) \end{aligned}$$

and hence, the minimum mean square estimate of f can be written as

$$f(x) = \frac{w(x)}{w_{BL}(x)} s_{BL}(x) \tag{3.35}$$

From this equation, it is easy to compute f given s_{BL} and w. All that is required is a DFT to obtain w_{BL} from w, which can be computed using an FFT. Furthermore, because the orthogonality principle can be applied to two-dimensional problems, the equations listed above may be used to extrapolate the spectrum of a two-dimensional band-limited signal.

3.13 Application of the maximum-entropy criterion

The application of the maximum-entropy criterion for solving the spectral extrapolation problem is usually attributed to Burg [1]. This technique is sometimes called the 'all-pole' method because of the nature of the estimation model used. It is a method often associated with the reconstruction of a power spectrum, but it can in fact be applied to any problem involving the reconstruction of signals from band-limited data.

Basic problem Given

$$F_n \equiv F(k_n) = \int_{-X}^{X} f(x) \exp(-ik_n x)\, \mathrm{d}x$$

where $|k_n| \leq K$ (the bandwidth), recover $f(x)$. This problem is equivalent to extrapolating the data F_n beyond the bandwidth K.

Solution A model for $f(x)$ is

$$f(x) = \frac{1}{|\sum_n a_n \phi_n(x)|^2} \tag{3.36}$$

where

$$\phi_n(x) = \exp(-ik_n x) \qquad \text{and} \qquad \sum_n \equiv \sum_{n=0}^{N-1}$$

The criterion for computing the a_n is:

the entropy E of the signal should be maximized

where E is defined as

$$E = \int_{-X}^{X} \ln f(x)\, \mathrm{d}x$$

This definition requires that $f(x) > 0\ \forall\ x$. The entropy measure $E(a_m)$ is a maximum when

$$\frac{\partial E}{\partial a_m} = 0 \qquad \text{for} \qquad m > 0$$

Using Eqn 3.36 for $f(x)$, the entropy can be written in the form:

$$E = -\int_{-X}^{X} \ln \left| \sum_n a_n \phi_n(x) \right|^2 \mathrm{d}x$$

Differentiating, we get

$$\begin{aligned}
\frac{\partial E}{\partial a_m} &= -\frac{\partial}{\partial a_m} \int_{-X}^{X} \ln \left| \sum_n a_n \phi_n(x) \right|^2 \mathrm{d}x \\
&= -\int_{-X}^{X} \frac{1}{|\sum_n a_n \phi_n(x)|^2} \frac{\partial}{\partial a_m} \left| \sum_n a_n \phi_n(x) \right|^2 \mathrm{d}x \\
&= \int_{-X}^{X} f(x) \sum_n a_n \phi_n(x) \phi_n^*(x)\, \mathrm{d}x = 0
\end{aligned}$$

Now using $\phi_n(x) = \exp(-ik_n x)$ we obtain

$$\sum_n a_n \int_{-X}^{X} f(x) \exp[-i(k_n - k_m)x] \, \mathrm{d}x = 0$$

or

$$\sum_n a_n F(k_n - k_m) = 0, \qquad m > 0$$

where the $F(k_n)$ are the given data. This is a 'data-consistent' result. In the case $m = 0$ we use the normalization condition

$$\sum_n a_n F(k_n - k_m) = 1, \qquad m = 0$$

We are then required to solve the following system of equations

$$\sum_{n=0}^{N-1} a_n F_{n-m} = \begin{cases} 1, & m = 0 \\ 0, & m > 0 \end{cases}$$

where $F_{n-m} = F(k_n - k_m)$. In matrix form, this system can be written as:

$$\begin{pmatrix} F_0 & F_1 & F_2 & \dots & F_{N-1} \\ F_{-1} & F_0 & F_1 & \dots & F_{N-2} \\ F_{-2} & F_{-1} & F_0 & \dots & F_{N-3} \\ \vdots & \vdots & \vdots & \ddots & \vdots \\ F_{1-N} & F_{2-N} & F_{3-N} & \dots & F_0 \end{pmatrix} \begin{pmatrix} a_0 \\ a_1 \\ a_2 \\ \vdots \\ a_{N-1} \end{pmatrix} = \begin{pmatrix} 1 \\ 0 \\ 0 \\ \vdots \\ 0 \end{pmatrix}$$

The characteristic matrix is a Toeplitz matrix. By solving this system of equations for a_n we can compute $f(x)$ using

$$f(x) = \frac{1}{\left| \sum_{n=0}^{N-1} a_n \exp(-ik_n x) \right|^2}. \tag{3.37}$$

REFERENCE

[1] J. P. Burg. Maximum entropy spectral analysis. Ph.D. thesis, Department of Geophysics, Stanford University, 1975.

4 Introduction to fractal geometry

In this chapter we investigate the use of fractal geometry for segmenting digital signals and images. A method of texture segmentation is introduced that is based on the fractal dimension. Using this approach, variations in texture across a signal or image can be characterized in terms of variations in the fractal dimension. By analysing the spatial fluctuations in-fractal dimension obtained using a conventional moving-window approach, a digital signal or image can be texture segmented; this is the principle of fractal-dimension segmentation (FDS). In this book, we apply this form of texture segmentation to isolated speech signals.

An overview of methods for computing the fractal dimension is presented, focusing on an approach that makes use of the characteristic power spectral density function (PSDF) of a random scaling fractal (RSF) signal. A more general model for the PSDF of a stochastic signal is then introduced and discussed with reference to texture segmentation.

We shall apply fractal-dimension segmentation to a number of different speech signals and discuss the results for isolated words and the components (e.g. fricatives) from which these words are composed. In particular, it will be shown that by pre-filtering speech signals with a low-pass filter of the form $1/k$, they can be classified into fractal dimensions that lie within the correct range, i.e. $[1, 2]$. This provides confidence in the approach to speech segmentation considered in this book and, in principle, allows a template-matching scheme to be designed that is based exclusively on FDS.

In this chapter, the segmentation of both signals and images is considered in general. Although the application described is for speech signals alone, the segmentation of images is also discussed as a prelude to an additional route to speech segmentation, namely, the FDS of time–frequency maps.

4.1 History of fractal geometry

Central to fractal geometry is the concept of *self-similarity*, in which an object appears to look similar at different scales; a fairly obvious concept when observing certain naturally occurring features, but one that has only relatively recently started to be

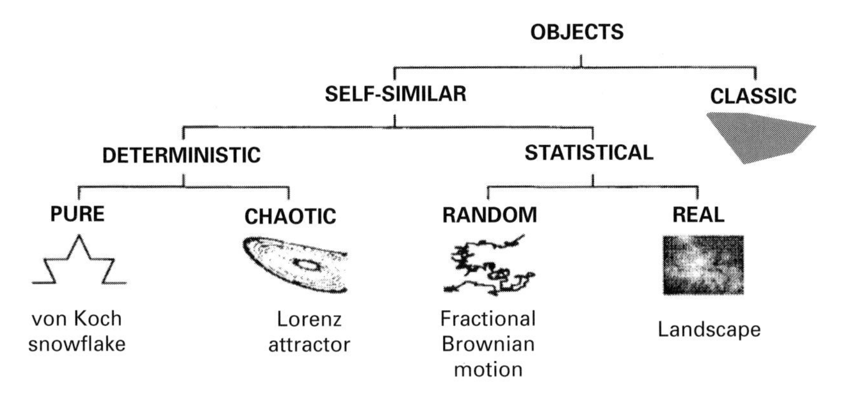

Figure 4.1 A simple typology of geometric objects. 'Classic' objects are those that are not self-similar.

developed mathematically and applied to various branches of science and engineering. This concept can be applied to systems of varying physical size, depending on the complexity and diversity of the fractal model that is considered. Ultimately, it is of philosophical interest to view the universe itself as a single fractal, the self-similar parts of which have yet to be fully categorized; those naturally occurring objects for which fractal models abound are considered as smaller subsets of a larger whole. This view is closely related to the concept of a chaotic universe in which the dynamical behaviour of a system cannot necessarily be predetermined. Such systems exhibit self-similarity when visualized and analysed in an appropriate way (i.e. in an appropriate phase space). In this sense, the geometry of a chaotic system may be considered to be fractal.

Self-similarity is a very general term. There are two distinct types of self-similarity (as illustrated in Figure 4.1):

- *deterministic self-similarity*, in which the fractal is composed of distinct features that resemble each other in some way at different scales (feature scale invariance)
- *statistical self-similarity*, in which the features of the fractal may change at different scales but the statistical properties at all scales are the same (statistical scale invariance)

Deterministic fractals are usually generated through some iterated-function system (IFS) and are remarkable for the complexity that can be derived through the simplest of these iterated systems. The way in which the output from these systems is viewed graphically and interpreted geometrically changes substantially from one fractal to another, but the overall principle remains the same.

Statistically self-similar fractals are used to model a variety of naturally occurring objects (background noise, clouds, landscapes, coastlines etc.). They can be generated through a variety of different stochastic modelling techniques. They can also be considered to be the solutions of certain classes of stochastic differential equations of fractional order, as discussed later in this chapter.

Table 4.1 *Fractal types and associated ranges of the fractal dimension D*

Fractal dimension	Fractal type
$0 < D < 1$	fractal dust
$1 < D < 2$	fractal signals
$2 < D < 3$	fractal surfaces (images)
$3 < D < 4$	fractal volumes
$4 < D < 5$	fractal time

The measure most commonly associated with a self-similar object is its *fractal* (or similarity) *dimension*. If we consider a bounded set A in a Euclidean n-dimensional space, then the set A is said to be self-similar if A is the union of N distinct (non-overlapping) copies of itself, each of which has been scaled by a ratio $r < 1$ in all coordinates. The fractal is described by the relationship (see [1] for example)

$$Nr^D = 1 \quad \text{or} \quad D = -\frac{\ln N}{\ln r} \tag{4.1}$$

where D is the fractal dimension. The ranges in the value of D characterize the type of fractal; see Table 4.1.

In each case, the fractal may be deterministic or random. A random fractal is taken to be composed of N distinct subsets, each of which is scaled down by a ratio $r < 1$ from the original and is the same in all statistical respects to the scaled original. The fractal dimension in this case is also given by Eqn 4.1. The scaling ratios need not be the same for all the scaled-down copies. Certain fractal sets are composed of the union of N distinct subsets each of which is scaled down by a ratio r_i, $1 \leq i \leq N$, from the original in all coordinates. The fractal dimension is given by a generalization of Eqn 4.1, namely

$$\sum_{i=1}^{N} r_i^D = 1$$

Finally, there are *self-affine* fractal sets, which are scaled by different ratios in different coordinates and which are not, in general, self-similar. For example, consider a curve $f(x)$ that satisfies

$$f(\lambda x) = \lambda^\alpha f(x) \qquad \forall \ \lambda > 0$$

where λ is a scaling factor and α is the scaling exponent. This equation implies that a scaling of the x-coordinate by λ gives a scaling of the f-coordinate by a factor λ^α; this is an example of self-affinity. A special case occurs when $\alpha = 1$, when we have a scaling of x by λ producing a scaling of f by λ; this is an example of self-similarity. Random fractal signals and images are, in general, examples of self-affine records.

Naturally occurring fractals also differ from strictly mathematically defined fractals in that they do not display statistical or exact self-similarity over all scales. Rather, they display fractal properties over a limited range of scales. Yokoya *et al.* [2] developed an algorithm for computing the upper and lower scale limits.

4.1.1 Self-similarity and self-affinity

Texture is a word that is commonly used in a variety of contexts but is at best a qualitative description of a sensation. Visual texture can be associated with a wide range of signals and images but the term cannot be taken to quantify any particular characteristic. How then can we quantify texture mathematically – is there a specific and unique definition for texture? To begin with, we can state that textural information is not well defined in terms of Euclidean geometry. Most objects can be divided into regions containing either

- deterministic information (for which Euclidean geometry is usually applicable), or
- textural information (which is not easily described by Euclidean geometry).

The latter case requires suitable methods of texture segmentation to be researched. Typically one or more 'measures' for texture can be defined and a moving window applied to the data. For each window position, each of the texture measures is computed. If only one measure is defined then one 'measure image' is produced, to which conventional digital signal processing can then be applied. If more than one measure is defined, then some sort of clustering algorithm is needed to reduce the several measure images into one final image. In addition, a scaling can be applied whereby the window size is gradually decreased, giving rise to yet more measure images at different scales.

It is worth stating at the outset that there is no universally accepted mathematical definition of texture. Mandelbrot [3] states the following:

- 'texture is an elusive notion which mathematicians and scientists tend to avoid because they cannot grasp it';
- 'much of fractal geometry could pass as an implicit study of texture'.

It is essential for the field of speech processing and computer vision that an unsupervised, automatic, means of extracting textural features that agrees with human sensory perception can be found. This is particularly important in the area of speech processing, where an analysis of speech signals often involves having to differentiate between contrasts in texture (e.g. background noise versus speech waveforms). More and more data is becoming available from an increasingly diverse range of applications, the rate of data acquisition often being vastly greater than the speed at which conventional analysis and interpretation can be performed. In the area of speech processing and speech synthesis, an increasing amount of research is being undertaken into the rate of analysis of on-line data in relation to the design of new instrumentation, for example. Most of this effort is concerned with the design of recognition systems in which features of particular phonetic significance can be detected automatically.

Although this book is concerned with the application of fractal geometry to texture segmentation, in which the fractal dimension is taken to be an appropriate measure, it is worth discussing other measures that have been considered. In [4], Tamura *et al.* define *six mathematical measures* that are considered necessary to classify a given texture. These measures are as follows.

1. *Coarseness*: coarse versus fine.

   Coarseness is the most fundamental textural feature and has been investigated thoroughly since the early studies [5, 6]. Often, in a narrow sense, coarseness is synonymous with texture. When two patterns differ only in scale, the magnified one is coarser. For patterns with different structures, the bigger the element size and/or the less the elements are repeated, the coarser a pattern is held to be.

2. *Contrast*: high contrast versus low contrast.

   The simplest method of varying signal or image contrast is by stretching or shrinking its amplitude or grey scale respectively [7]. By changing the contrast of an image we alter the image quality, not the image structure. When two patterns differ only in grey-level distribution, the difference in their contrast can be measured. However, we may suppose that more factors influence the contrast difference between two texture patterns with different structures. The following four factors are considered for a definition of contrast:

   (i) the dynamic range;
   (ii) the polarization of the distribution of dark-field and bright-field regions in the grey-level histogram, i.e. the ratio of dark and bright areas;
   (iii) the sharpness of edges (images with sharp edges have higher contrast);
   (iv) the period of repetition of patterns.

3. *Directionality*: directional versus non-directional.

   This is a global property over the region. Directionality involves both the element shape and the placement rule. Bajcsy [8, 9] divides directionality into two groups, monodirectional and bidirectional. If only the total degree of directionality is considered then the orientation of the texture pattern does not matter, i.e. patterns that differ only in orientation should have the same degree of directionality.

4. *Line-likeness*: line-like versus blob-like.

   This concept is concerned with the shape of a texture element. It is expected that this feature should supplement the major ones previously mentioned, especially when two patterns cannot be distinguished by directionality.

5. *Regularity*: regular versus irregular.

   This is a property concerning variations in a placement rule. However, it can be supposed that any variation in the elements, especially in the case of natural textures, reduces the regularity on the whole. Additionally, a fine structure tends to be perceived as regular.

6. *Roughness*: rough versus smooth.

   This description was originally intended for tactile textures, not for visual textures.

However, when we observe natural textures we are usually able to compare them in terms of rough or smooth. It is still a matter for debate as to whether this subjective judgement is due to the total energy of changes in grey level or to our imaginary tactile sense.

Tamura *et al.* in [4] give mathematical definitions of all these measures, together with the results of experiments to find the correlation between the measures and human perception. The work reported in this book is an approach to the problem of image discrimination and classification through texture that is an alternative to the other principal approaches, which are:

- the statistical co-occurrence approach (see [10], for example)
- the method of mathematical morphology
- scale space filtering (see [11], for example)

It should be noted, however, that these approaches are not mutually exclusive. For example, some recent papers on morphology also include fractal ideas ([12], for example) and in [13] the performances of four classes of textural features are compared with the conclusion that

> the results show that co-occurrence features perform best followed by fractal features . . . . However, there is no universally best subset of features. The feature selection task has to be performed for each specific problem to decide which feature of which type one should use.

In this book, the use of the fractal dimension as a measure for texture is discussed. Different methods of computing this measure for signals and images are introduced and the theoretical basis for this approach to texture segmentation is presented. A new texture model is also proposed which is based on a generalization of the PSDF of a random scaling fractal (RSF) that incorporates stochastic processes such as the Ornstein–Uhlenbeck process and the Bermann process.

4.1.2 Random scaling fractals

Given that natural surfaces can be approximated as fractals over a range of scales, it is necessary to examine how the imaging process maps fractal surfaces into grey-level surfaces. Kube and Pentland [14] showed that the image of a fractal surface is itself a fractal surface under the following conditions:

- the surface is Lambertian;
- the surface is illuminated by (possibly several) distant point sources;
- the surface is not self-shadowing.

This book is concerned with the segmentation of signals and images through the computation of fractal dimensions that give a measure of the 'roughness' or texture. A high-value fractal dimension indicates a rough signal or surface whereas a low value indicates a smooth signal or surface. A variety of methods have been developed to calculate the fractal dimension of signals and images. However, the large majority of the literature concentrates on computer vision and image processing problems. There is significantly less published material concerning signal processing and in particular speech processing. The following section provides an overview of those techniques that have been developed to compute the fractal dimension of a signal and/or image.

A random scaling fractal (RSF) is one that exhibits statistical self-affinity, i.e. an RSF signal $f(x)$ has the following characteristic property:

$$\Pr[f_\lambda(x)] = \lambda^{-\alpha} \Pr[f(\lambda x)] \tag{4.2}$$

where $\Pr[\cdot]$ denotes the probability density function. The interpretation of this result is that as we zoom in to observe the detailed structure of a signal, changing the scale by a factor of λ, the distribution of amplitudes remains the same (subject to scaling by λ^α) even though the signal itself 'looks' different at different scales, i.e. $f_\lambda(x) \neq f(\lambda x)$. When $\alpha = 1$ and

$$\Pr[f_\lambda(x)] = \lambda \Pr[f(\lambda x)]$$

we obtain a statistically self-similar signal.

The purpose of this section is to classify RSF signals and images in terms of an appropriate (Fourier-based) theory applied to the solution of certain classes of stochastic differential equations of fractional order. In addition to leading directly to a power spectrum method for computing the fractal dimension, this approach leads to a more general classification of stochastic signals, of which RSF signals are a special case.

Stochastic differential equations of fractional order

Consider the following stochastic fractional differential (SFD) equation for a signal $f(x)$:

$$\frac{\mathrm{d}^q}{\mathrm{d}x^q} f(x) = n(x) \tag{4.3}$$

where $q > 0$ and n is white noise (i.e. noise whose PSDF is a constant). Note, given that n is dimensionless, when $q = 1$ the solution to this equation has dimension 1 (x being taken to be a length) and when $q = 2$ the solution has dimension 2. The solution to this equation for $1 < q < 2$ can therefore be regarded in terms of a fractal signal. The relationship between q – the 'Fourier dimension' – and the fractal dimension D is discussed later (see subsection 4.3.4).

The aim of the following discussion is to show that the solution to this equation is one that is consistent with Eqn 4.2. As a consequence of this, we are able to quantify the PSDF of a fractal signal.

Consideration must be given first to the method of solution required to solve Eqn 4.3. The subject of fractional calculus is not new [15, 16], but its use in defining RSF signals and stochastic processes in general has not been fully realized. There are many approaches to working with fractional derivatives but they all rely on the generalization of results associated with derivatives of integer order. A definition that is particularly useful is

$$\frac{d^q}{dx^q} f(x) = \frac{1}{2\pi} \int_{-\infty}^{\infty} (ik)^q F(k) \exp(ikx) \, dk$$

where $F(k)$ is the Fourier transform of $f(x)$, given by

$$F(k) = \int_{-\infty}^{\infty} f(x) \exp(-ikx) \, dx$$

Here, k is the spatial frequency. Using this definition, the solution to Eqn 4.3 becomes

$$f(x) = \frac{1}{2\pi} \int_{-\infty}^{\infty} \frac{1}{(ik)^q} N(k) \exp(ikx) \, dk$$

where $N(k)$ is the Fourier transform of $n(x)$. Here, the solution is expressed in terms of the inverse Fourier transform of the product of two functions, i.e. $(ik)^{-q}$ and $N(k)$. Application of the convolution theorem allows us to write this result in the form

$$f(x) = \int h(x - y) n(y) \, dy$$

where h is given by

$$h(x) = \frac{1}{2\pi} \int_{-\infty}^{\infty} \frac{\exp(ikx)}{(ik)^q} \, dk$$

Substituting p for ik, $h(x)$ can be written in terms of the inverse Laplace transform of p^{-q}. Since

$$\hat{L}[x^q] = \frac{\Gamma(q+1)}{p^{q+1}}, \qquad q > -1, \quad \mathrm{Re}[p] > 0$$

where $\hat{L}$ is taken to denote the Laplace transform and Γ is the Gamma function,

$$\Gamma(q) = \int_{0}^{\infty} t^{q-1} \exp(-t) \, dt,$$

we can write

$$x^q = \hat{L}^{-1} \left[\frac{\Gamma(q+1)}{p^{q+1}} \right]$$

or

$$\hat{L}^{-1}\left[\frac{1}{p^q}\right] = \frac{1}{\Gamma(q)} x^{q-1}$$

Thus, the solution to Eqn 4.2 can be written in the form

$$f(x) = \frac{1}{\Gamma(q)} \int_0^x \frac{n(y)}{(x-y)^{1-q}} \, dy$$

This is the Liouville–Riemann transform and is an example of a fractional integral. Is this transform consistent with the concept of statistical self-affinity? Consider

$$f'(x) = \frac{1}{\Gamma(q)} \int_0^x \frac{n(\lambda y)}{(x-y)^{1-q}} \, dy$$

where λ is a scaling parameter. Substituting $z = \lambda y$, we obtain

$$f_\lambda(x) = \frac{1}{\lambda^q} \frac{1}{\Gamma(q)} \int_0^{\lambda x} \frac{n(z)}{(\lambda x - z)^{1-q}} \, dz = \frac{1}{\lambda^q} f(\lambda x)$$

Now both $f_\lambda(x)$ and $f(\lambda x)$ are stochastic functions of the same type but over different scales ($n(x)$ being the white noise at any scale) and so, although $f_\lambda(x) \neq f(\lambda x)$, we have

$$\Pr[f_\lambda(x)] = \frac{1}{\lambda^q} \Pr[f(\lambda x)]$$

which describes a statistically self-affine signal.

The relationship between white noise and fractal noise can be considered in terms of fractional differentiation and fractional integration, as illustrated in Figure 4.2. Note that the normal interpretation of integration as being a 'smoothing' process and differentiation as being a 'roughening' process, for integer order, continues to hold for fractional order.

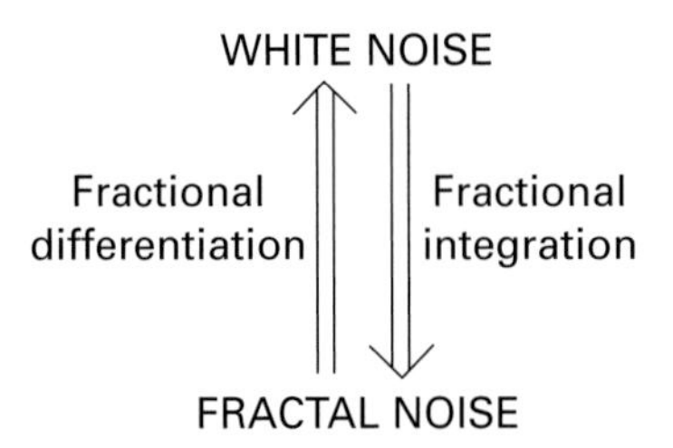

Figure 4.2 The relationship between fractal and white noise.

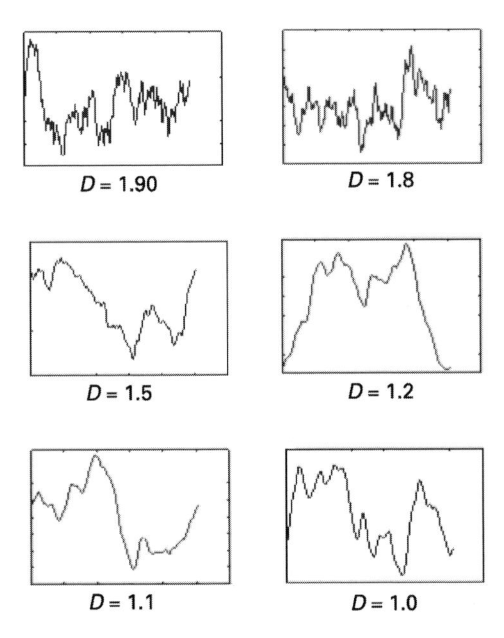

Figure 4.3 The signal $f(x)$ for fractal noise with various values of fractal dimension D.

Foreward algorithm

The algorithm used to obtain $f(x)$ from the fractal dimension D is as follows.

Step 1 Compute a random white Gaussian n_i, $i = 1, 2, \ldots, N$.

Step 2 Fourier-transform n_i to create $N_i(\omega)$.

Step 3 Filter in complex space to get $F(\omega) = (i\omega)^{-\beta}/N_i(\omega)$, where $\beta = 2q$ (see subsection 4.3.1).

Step 4 Inverse-Fourier-transform $F(\omega)$ to create $f(x) = \text{Re}[F^{-1}F(\omega)]$. The results obtained for different values of $D = (5 - \beta)/2 = (5 - 2q)/2$ are shown in Figure 4.3.

4.2 Fractal-dimension segmentation

4.2.1 Fourier dimension and fractal dimension

As with many other techniques of digital signal and image processing, computation of the fractal dimension can be undertaken in ‘real space’ (processing the data directly) or in ‘transform space’ (processing the data after taking an appropriate integral transform). In the latter case, use can be made of the Fourier transform, as the PSDF of a fractal signal or image has the expected form. This important relationship between a random scaling fractal and its PSDF is introduced later, where it forms the basis for a more general discussion on stochastic modelling using PSDF models.

In general, there is no unique and general rule for computing the fractal dimension. A large number of algorithms have been developed over the past 15 years for computation of the fractal dimension. These can be broadly categorized into two families.

So far, in this book, models for an RSF field have been expressed in terms of the Fourier dimension q alone. In addition, the appropriate ranges that should be attributed to the Fourier dimension for a fractal signal and for a fractal or Mandelbrot surface have been left ambiguous. In this section, the relationship between the fractal dimension D and the Fourier dimension q is quantified. The basic result is derived in [17], where it is shown that the Fourier dimension q and the box-counting dimension D (the fractal dimension) for a fractal signal are related as follows:

$$q \leq D + \tfrac{1}{2}$$

In addition, we note that the relationship between the fractal dimension of a fractal surface or image, D_2, and the fractal dimension of a fractal signal, D_1, is given by [18]

$$D_2 = 1 + D_1$$

With these basic results, it is then trivial to determine the ranges of the Fourier dimensions for a fractal signal and a fractal image.

A mathematical model for fractal-dimension segmentation

We can consider a non-stationary model for a fractal signal $f(x)$ given by

$$\frac{\mathrm{d}^{q(x)}}{\mathrm{d}x^{q(x)}} f(x) = n(x)$$

where $q(x)$ is *quantized*, i.e.

$$q(x) = \begin{cases} q_1, & x_1 \leq x < x_2 \\ q_2, & x_2 \leq x < x_3 \\ \vdots & \vdots \end{cases}$$

The interval $x_{i+1} - x_i$ describes the window over which the fractal signal is assumed to be stationary, with Fourier dimension q_i. Here, we assume that the window is of a fixed length. Fractal-dimension segmentation can then be considered in terms of the inverse problem:

given $f(x)$ evaluate $q(x)$.

The methods of solution use one of the following:

- size–measure relationships, based on recursive length or area measurements of a curve or surface using different measuring scales
- relationships based on approximating or fitting a curve or surface to a known fractal function or to a statistical property such as the variance

By way of a brief introduction to method 1 above, consider a simple Euclidean straight line ℓ of length $L(\ell)$ over which we 'walk' a shorter ruler of length δ. The number of steps taken to cover the line $N[L(\ell), \delta]$ is then L/δ, which is not always an integer for arbitrary L and δ. Hence, we should compute $N(L, \delta)$ and $N(L, \delta) + 1$, but as $\delta \to 0$ the errors become small. Clearly

$$N[L(\ell), \delta] = \frac{L(\ell)}{\delta} = L(\ell)\delta^{-1}$$

This implies that

$$1 = \frac{\ln L(\ell) - \ln N[L(\ell), \delta]}{\ln \delta} = -\frac{\partial \ln N[L(\ell), \delta]}{\partial \ln \delta}$$

This ratio expresses the topological dimension $D_T = 1$ of the line. Here $L(\ell)$ is the Lebesgue measure of the line and, if we normalize by setting $L(\ell) = 1$, the latter equation can then be written as

$$1 = -\lim_{\delta \to 0} \left[\frac{\ln N(\delta)}{\ln \delta} \right]$$

since the smaller is δ, the smaller is the error in counting $N(\delta)$. We also then have $N(\delta) = \delta^{-1}$.

For extension to a fractal curve f, the essential point is that the various fractal dimensions satisfy an equation of the form

$$N[F(f), \delta] = F(f)\delta^{-D}$$

where $N[F(f), \delta]$ is read as the number of rulers of size δ needed to cover a fractal set f whose measure is $F(f)$, which can be any suitable valid measure of the curve. Again we may normalize, which amounts to defining a new measure F' as some constant times the old measure, and get

$$D = -\lim_{\delta \to 0} \left[\frac{\ln N(\delta)}{\ln \delta} \right]$$

where $N[F'(f), \delta]$ is written as $N(\delta)$ for notational convenience. In practice, if we are dealing with a digital signal (a sampled curve), rather than an abstract analogue mathematical curve that has precise fractal properties over all scales, then

$$D = -\left\langle \frac{\partial \ln N(\delta)}{\partial \ln \delta} \right\rangle \tag{4.4}$$

where we choose values δ_1 and δ_2 satisfying $\delta_1 < \delta < \delta_2$ over which we do some sort of averaging process denoted by $\langle\ \rangle$. The most common approach is to look at the bilogarithmic graph of $\ln N(\delta)$ against $\ln \delta$, choose values δ_1 and δ_2 over which the graph appears to be straight and then apply a least squares fit to the straight line within these limits.

The least squares approximation

All algorithms discussed in this section use a least squares approach to computing the fractal dimension. It is therefore worth briefly reviewing this technique.

Let $f_i, i = 1, 2, \ldots, N$, be a real digital function consisting of N elements and let $\hat{f}_i$ be an approximation to this function; we assume that $\hat{f}_i$ is the expected form of the data f_i. The least squares error e is then defined as

$$e = \sum_i (f_i - \hat{f}_i)^2$$

In most cases, algorithms for computing the fractal dimension use logarithmic or semi-logarithmic plots to fit the results of a given algorithm to a line. In these cases, we are interested in finding the slope β, and sometimes the intercept c, of the line

$$\hat{f}_i = \beta x_i + c$$

To find the best fit, we minimize the error e, which is taken to be a function of β and c. This is achieved by finding the solutions to the equations

$$\frac{\partial e}{\partial \beta} = 0 \quad \text{and} \quad \frac{\partial e}{\partial c} = 0$$

Differentiating with respect to β and c gives

$$\sum_i x_i(f_i - \beta x_i - c) = 0$$

and

$$\sum_i (f_i - \beta x_i - c) = 0$$

Solving for β and c we obtain

$$\beta = \frac{N \sum_i f_i x_i - \left(\sum_i f_i\right)\left(\sum_i x_i\right)}{N \sum x_i^2 - \left(\sum_i x_i\right)^2}$$

and

$$c = \frac{\sum_i f_i - \beta \sum_i x_i}{N}$$

This approach can also be used when the data is two-dimensional (a digital image or grey-level surface), so that we approximate the data f_{ij} by a function

$$\hat{f}_{ij} = \beta x_{ij} + c$$

The result (i.e. the expression for β and c) is the same as above except that the summation is over i and j. In the following sections, algorithms for computing the data used to

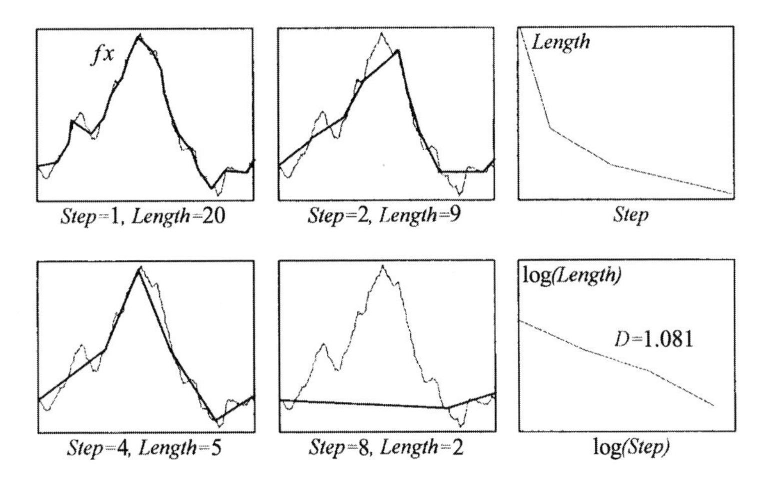

Figure 4.4 Illustration of the walking-divider method for computing the fractal dimension D of a signal $f(x)$, showing four iterations (left-hand and middle columns) and the least squares fit (right-hand columns).

calculate the fractal dimension by the least squares method are discussed. Some of these algorithms are based on the following relationship

$$Length = c \times Step^{\beta}$$

which can be linearized thus:

$$\ln(Length) = \ln c + \beta \ln(Step)$$

Here, *Length* represents the measurement value of the curve or surface obtained using a ruler of size $Step$. The slope of the log–log plot is β, which has a simple algebraic relationship with the fractal dimension D, depending on the algorithm used.

The walking-divider method Introduced by Shelberg [19], this method uses a chord length ($Step$) and measures the number of chord lengths ($Length$) needed to cover a fractal curve, as explained above. The technique is based on the principle of taking rulers of smaller and smaller size $Step$ to cover the curve and counting the number of rulers $Length$ required in each case. This approach is based on a direct interpretation of Eqn 4.4, in which $N(\delta) \equiv Length$ and $\delta \equiv Step$ are estimated in a systematic fashion. It is a recursive process in which the value of $Step$ is repeatedly decreased (typically halved) and the new value of $Length$ calculated. Here, the input signals are taken to be of size N, where N is a power of 2, because of the recursive nature of the method. A least squares fit to a bilogarithmic plot of $Length$ against $Step$ gives β, where $D = -\beta$. This part of the calculation essentially provides an estimate of the average gradient in Eqn 4.4, as illustrated in Figure 4.4.

The walking-divider method suffers from a number of problems. First of all, the initial value of $Step$ must be carefully chosen. Shelberg [19] and others [20, 21]

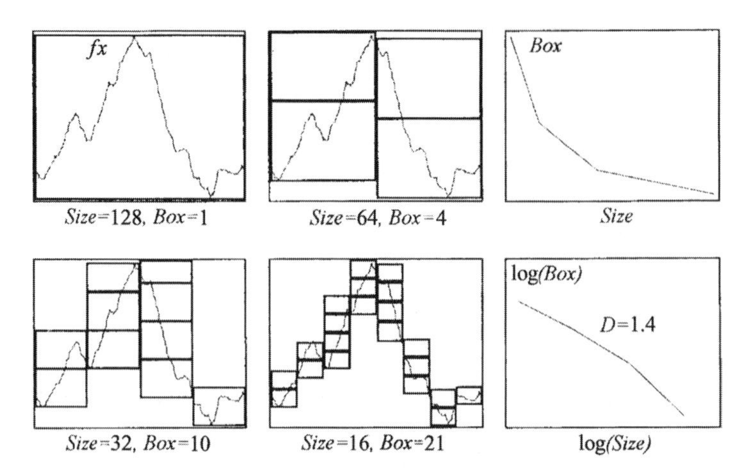

Figure 4.5 Illustration of the box-counting method for computing the fractal dimension D of a signal $f(x)$, showing four iterations and the least squares fit $D = -\beta$.

describe an appropriate starting value as half the average distance between the points. Clearly, computation of this initial value, and the procedure required to count *Length*, makes this algorithm time consuming.

The box-counting method One of the most popular algorithms for computing the fractal dimension of fractal signals and images is the box-counting method, originally developed by Voss [22] but modified by others to achieve a reasonably fast and accurate algorithm. The numerous algorithms published which are based on this theme follow the same basic principle.

Box counting in general involves covering a fractal with a grid of n-dimensional boxes, or hypercubes, with side length δ and counting the number of non-empty boxes $N(\delta)$. For signals, the grid is one of squares and, for images, of cubes. Boxes of recursively different sizes are used to cover the fractal. Here again, an input signal with N elements or an image of size $N \times N$, where N is a power of 2, is used as input. The slope β obtained in a bilogarithmic plot of the number of boxes used, *Box*, against their side length, *Size*, then gives the fractal dimension, also known as the box or Minkowski dimension; see Figure 4.5 successive divisions by a factor 2 are used for *Size* to give a regular spacing in the bilogarithmic plot and least squares fit. In practice, a regular grid is usually applied (see Figure 4.6) to the data and the non-empty boxes are counted.

The behaviour of this algorithm is such that the greater the number of points used for the least squares fit, the better the estimate of the fractal dimension. In a two-dimensional version of this algorithm, Sarkar and Chaudhuri [23] and Sarkar, Chaudhuri and Kundu [24] considered the problem of optimizing the number of boxes, given the *Size* value, required to compute an accurate fractal dimension. The solution is to map the entire image with boxes (cubes) and then identify the lowest and highest possible values of *Box* needed to cover the surface.

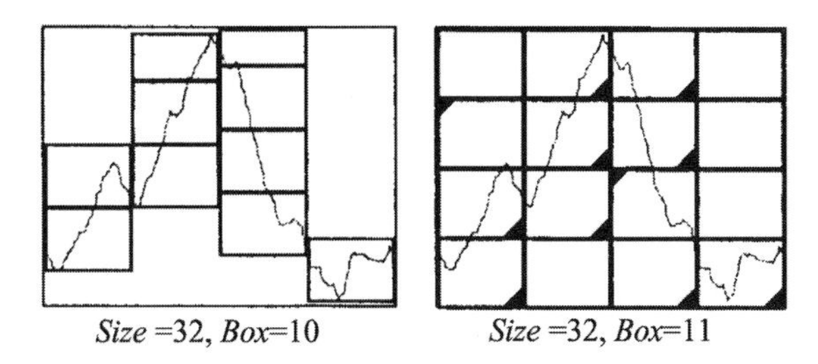

Figure 4.6 Illustration of irregular and regular grids used for the box-counting algorithm; for the regular grid the non-empty boxes are indicated by the black corner triangles.

In general, box-counting algorithms behave well and produce accurate estimates for fractal dimensions between 1 and 1.5 for digital signals and between 2 and 2.5 for digital images and in these ranges are easy to code and fast to compute. For higher fractal dimensions they tend to give less accurate results, underestimating in most cases and saturating near 0.6 above the topological dimension D_T, as confirmed by Keller and Chen [25].

The prism method Clarke [26] defined an algorithm, based on the idea of box counting: instead of counting the number of boxes in a region for a given box size, the area based on four triangles defined by the corner points is computed and summed over a grey-level surface. The triangles define a prism based on the elevated corners and a central point computed in terms of the average of the four corners. A bilogarithmic plot of the sum of the prisms' areas for a given base area gives a fit to a line whose slope is β, in which $D = 2 - \beta$. The basic engine for this algorithm is similar to the box-counting method but is slower, owing to the number of multiplications implied by the calculation of the areas.

Hybrid methods

Hybrid methods calculate the fractal dimension of two-dimensional surfaces using one-dimensional methods. This approach is based on a result by Goodchild [18], who confirmed a simple relationship between the fractal dimensions of a surface's contours (one-dimensional fractal curves) and the fractal dimension of the surface itself, namely the relation mentioned at the start of this section,

$$D_2 = 1 + D_1$$

where D_1 is the average fractal dimension of the contour lines and D_2 is the fractal dimension of the surface. In principle, this result holds for any algorithm used to compute D_1.

Contour lines As a direct consequence of this fundamental result, Shelberg *et al.* [27] developed an algorithm based on extracting N lines of the same elevation in a grey-level surface, computing their fractal dimensions $D_{(1,n)}$ using the walking-divider technique and then finding the general two-dimensional fractal dimension D via the formula

$$D = 1 + \frac{1}{N} \sum_{n=1}^{N} D_{(1,n)}$$

This method was used with some success by Lam [28] to measure the fractal dimensions of Landsat TM images. The results were generally close to those computed using the box-counting method, for fractal dimensions between 2.1 and 2.4. However, to use this approach requires a certain amount of preprocessing [21, 29].

Robust fractal estimator This term was introduced by Clarke and Schweizer [20] as an alternative algorithm using the one-dimensional–two-dimensional method. Instead of considering elevation slices, profiles in the north–south and east–west directions are taken in order to compute the general fractal dimension. Each vertical intersection is processed via the walking-divider method and a new map of fractal dimensions created, where each point is defined by the average fractal dimension of the two profiles intersecting at that position.

Vertical-slice averaging This approach considers the fractal dimension of the surface to be the normal average of the fractal dimensions of all the vertical slices in the x and y directions plus 1. To add flexibility, this technique can be implemented with the possibility of computing in either direction or both directions (rows and/or columns) and of considering only a limited number of slices. Another choice can be made with regard to the one-dimensional algorithm used to compute the fractal dimensions.

Closed fractal curves For non-fractal closed curves in the plane, the perimeter ℓ is related to the enclosed area A by

$$\ell = c\sqrt{A}$$

where c is a constant for a given type of shape (e.g. for squares $c = 1$ and for circles $c = 2\sqrt{\pi}$. In [3] Mandelbrot generalized this equation to the case of closed fractal curves, to give

$$\ell = c\left(\sqrt{A}\right)^{D}, \quad 1 < D < 2$$

De Cola [30] used an algorithm based on this generalized perimeter–area relationship to classify different types of land-usage areas from Landsat Thematic Mapper SAR images. He also proposed a complete image-segmentation scheme using parallel processing.

Fractal properties and dimensions

Most authors have made the comment that the fractal dimension alone is not sufficient to quantify a given texture. Many quite different fractals can have the same value of fractal dimension, since

$$D = -\frac{\ln N}{\ln r} = -\frac{\ln N^a}{\ln r^a}$$

for all values of a provided that $r^a < 1$. For this reason, other fractal measures have been proposed to supplement the fractal dimension D in an attempt to define 'texture' uniquely.

The fractal signature Many methods of computing the fractal dimension depend on the use of *Richardson's equation* for some measured property M that is a function of scale ϵ,

$$M(\epsilon) = c\epsilon^{D_T - D}$$

where c is a constant and D_T is the topological dimension. A single value of D can then be computed using a bilogarithmic least squares fit. If, however, we use the fact that $M(1) = c$ we can compute D for $\epsilon = 2, 3, \ldots$. A plot of D against ϵ is called the *fractal signature.*

In reference [31] examples are given of the fractal signatures for different types of sample imagery. Natural imagery such as a dense tree background gives a slowly changing, essentially constant, signature. A synthetic image consisting of a single edge gives a low D at small ϵ rising to $D = 3$ at $\epsilon = 4$.

The correlation dimension and signature Each pixel in a grey-level image can be regarded as a point in a three-dimensional space $X_k = [i, j, g_{ij}]$, where i and j are the spatial coordinates of a pixel and g_{ij} is the grey level at these coordinates. For each pixel, a cube of size $2\epsilon + 1$ is constructed centred at the pixel. The number of points X_ℓ that fall inside this cube is counted for various values of ϵ. The probability $C(\epsilon)$ that at least one point lies within the cube can then be obtained by dividing the number of points by the cube volume,

$$C(\epsilon) = \frac{1}{N(2\epsilon + 1)^3} \sum_{k=1}^{N} \sum_{\substack{\ell=1 \\ \ell \neq k}} H(\epsilon - |X_k - X_\ell|)$$

where H is the Heaviside step function and N is the number of pixels in the image, if we are considering the whole image, or the number of pixels in a moving window and

$$H(\xi) = \begin{cases} 1, & \text{if } \xi \geq 0 \\ 0, & \text{if } \xi < 0 \end{cases}$$

The probability $C(\epsilon)$ obeys the *Richardson law* with a correlation dimension D_C (a special case of the correlation dimension ν defined later in this subsection):

$$C(\epsilon) = c(2\epsilon + 1)^{3 - D_C}$$

Here, a single value of D_C can be computed by the normal bilogarithmic least squares fit, or we may compute $C(0) = c$ and then compute a value of $C(\epsilon)$ for $\epsilon = 1, 2, \ldots$. In the latter case, we obtain the correlation signature as a plot of D_C against ϵ. For an image, the values of D_C range from 2 to 3. A highly correlated surface gives a correlation dimension close to 2 whereas a highly uncorrelated surface gives a value close to 3. In [31], examples of correlation signatures for a selection of real and synthetic images are given. In [32], the correlation dimension was used to search for cracks in the ice in NOAA images made by the Finnish Meteorological Institute. Results are also given of the application of this method to three-dimensional microscopy using a confocal light microscope.

Lacunarity 'Lacunarity' is another term due to Mandelbrot, from the Latin *lacuna*, meaning gap. The most straightforward illustration comes from considering the class of Cantor dusts, for which the fractal dimension, in this case, the similarity dimension, is given by $D = -\ln N/\ln r$ where N is the number of copies of the real line interval [0, 1] and $r < 1$ is the scaling factor. Clearly, there is an infinite set of pairs $\{N, r\}$ that give the same D. For example, the classic triadic Cantor set [2, 1/3] gives the same D as [4, 1/9], [8, 1/27] etc. These point sets will appear quite different, however. The differences lie in the way the gaps are distributed.

It was Mandelbrot [3] who first defined suitable mathematical measures for the lacunarity of deterministic fractal sets such as the Cantor set discussed above. Such definitions are unsuitable, however, for random fractals and he continued to give suitable candidate definitions of lacunarity for the class of random fractals. These definitions are based on the idea of mass distribution. Consider, for example, a curve $f(x)$. If the curve has negative values, then translate so that these are non-negative; now consider the values of f as representing mass so that we regard f as a mass distribution over the support of f. Mandelbrot proposed definitions such as

$$A = \left\langle \left(\frac{f}{\langle f \rangle} - 1 \right)^2 \right\rangle = \frac{\langle f^2 \rangle}{\langle f \rangle^2} - 1$$

and

$$A = \left\langle \left| \frac{f}{\langle f \rangle} - 1 \right| \right\rangle$$

for the lacunarity A, where $\langle \ \rangle$ denotes the mean. Note that the first definition can be expressed as the variance divided by the square of the mean. The definitions are not dissimilar to generalized statistical moments and this provides the key to multifractals, i.e. the higher-order fractal approach.

Higher-order fractals and dimensions

In principle, there is an unlimited number of fractal or generalized dimensions; this has given rise to the growing use of multifractals, i.e. higher-order fractals, in the mathematical and physical sciences. It is stated in the work of many authors on fractals that the fractal dimension alone is not sufficient to characterize a fractal set. The fractal dimension discussed in this book is only one of a number of generalized dimensions, as mentioned above. However, these generalized dimensions are defined via a measure-theoretic analysis in 'real space'. The use of transforms, and in particular the Fourier transform, applies principally to the similarity dimension (see below). This approach differentiates between regions of an image having different fractal dimensions but not between regions having the same fractal dimension but differing correlation dimension, information dimension and so on (see below).

A good example of the ineffectiveness of the fractal dimension follows from the result that the fractal dimension D of the sum of two fractal curves C_1 and C_2, with fractal dimensions D_1 and D_2, is given in [33] as

$$D = \max[D_1, D_2]$$

Suppose, for example, that D_1 is significantly larger than D_2; C_2 could be a square wave, for example, in which case $D_2 = 1$. Whilst it is clear to the eye that the merged curve is quite different from C_1 (or C_2), the same value of fractal dimension is found for the merged curve and for C_1.

The need for multifractal measures can easily be grasped by considering the definition and calculation of the box dimension. Consider the case of a point set with a large but finite number N of points embedded in fractal dimensional space and suppose that we cover the set with a uniform grid of hypercubes of size δ and count the number $M(\delta)$ of non-empty boxes. Such a strategy does not register information concerning the distribution of the number of points in the non-empty boxes. Suppose there are N_k points in the kth hypercube, and let $p_k = N_k/N$.

The *similarity dimension* (which takes account only of the total of the non-empty boxes $M(\delta)$ and not the N_k) is then given by

$$D = -\lim_{\delta\to 0}\lim_{N\to\infty}\frac{\ln M(\delta)}{\ln \delta}$$

The *information* (or *Renyi*) *dimension* σ is defined by

$$\sigma = -\lim_{\delta\to 0}\lim_{N\to\infty}\frac{S(\delta)}{\ln \delta}$$

where

$$S(\delta) = -\sum_{k=1}^{M(\delta)} p_k \ln p_k$$

The *correlation dimension* ν is defined in general as

$$\nu = \lim_{\delta \to 0} \lim_{N \to \infty} \frac{\ln C(\delta)}{\ln \delta}$$

where

$$C(\delta) = \frac{1}{N^2} \sum_{i \neq j} H(\delta - |X_i - X_j|)$$

and again H is the Heaviside step function. $C(\delta)$ counts the number of points whose distance $|X_i - X_j|$ is less than δ. In [34], D, σ and ν are the first three numbers of a hierarchy of generalized dimensions D_q for $q \geq 0$, i.e.

$$D = \lim_{q \to 0} D_q, \qquad \sigma = \lim_{q \to 1} D_q, \qquad \nu = \lim_{q \to 2} D_q$$

For $q = 3, 4, \ldots, n$ we have correlation dimensions associated with triplets, quadruplets and n-tuplets of points. The same authors show that the D_q form a non-increasing series:

$$D_q > D_{q'} \qquad \text{for any } q' > q$$

with the inequality replaced by an equality if and only if the fractal is homogeneous. Hence, we see that these various dimensions, previously introduced quite independently, form a general series. Moreover, a general expression for D_q was given by Falconer [33]:

$$D_q = \frac{1}{(q-1)} \lim_{\delta \to 0} \frac{\ln \left(\sum_i p_i^q \right)}{\ln \delta}$$

A valuable summary of the uses of both the fractal and chaos models (in which correlation dimensions are often computed) in imaging science is given in [35].

4.3 Non-stationary fractal signals

4.3.1 Inverse solution

The forward and inverse problems associated with fractal signals are to be posed in terms of the stochastic fractional differential equation (Eqn 4.3)

$$\frac{\mathrm{d}^q}{\mathrm{d}x^q} f(x) = n(x)$$

where $n(x)$ is the white noise; they can be stated as follows.

1. **Forward problem** Given q, compute f.
2. **Inverse problem** Given f, compute q.

The inverse problem can be approached in terms of the characteristic PSDF given by $P(k) = |F(k)|^2$, which is proportional to $|k|^{-2q}$, i.e.

$$P(k) = \frac{c}{|k|^\beta}, \qquad \beta = 2q$$

where c is a constant of proportionality. Note that c provides a measure of the 'energy' of the signal, since from Rayleigh's theorem

$$\text{energy} = \int_{-\infty}^{\infty} |f(x)|^2 \mathrm{d}x = \int_{-\infty}^{\infty} |F(k)|^2 \,\mathrm{d}k = c \int_{-\infty}^{\infty} \frac{1}{|k|^2} \,\mathrm{d}k$$

This result provides another way of computing β and hence the fractal dimension. This approach is based on using Fourier-transformed data, which in practice can be obtained using the fast Fourier transform (FFT) algorithm.

Consider a case in which the digital power spectrum $P_i \equiv P(k_i)$ is given, having been obtained by applying an FFT to the digital fractal signal $f_i, i = 1, 2, \ldots, N$ (N being a power of 2). This data can be approximated by

$$\hat{P}_i = \frac{c}{|k_i|^\beta}$$

and the least squares method used to obtain an estimate for β (and c).

Now consider the error function

$$e = \sum_i (\ln P_i - \ln \hat{P}_i)^2 = \sum_i [\ln P_i - (C - \beta \ln |k_i|)]^2$$

where $C = \ln c$ and it is assumed that $P_i > 0$ and $k_i > 0 \;\; \forall i$. Note that taking the difference between the logarithms of the data and of its approximated form linearizes the problem. The error e, which is a function of C and β, is a minimum when

$$\frac{\partial e}{\partial \beta} = 0 \qquad \text{and} \qquad \frac{\partial e}{\partial C} = 0$$

Solving these equations yields the following results:

$$\beta = \frac{N \sum_i (\ln P_i)(\ln |k_i|) - \left(\sum_i \ln |k_i|\right)\left(\sum_i \ln P_i\right)}{\left(\sum_i \ln |k_i|\right)^2 - N \sum_i (\ln |k_i|)^2}$$

and

$$C = \frac{1}{N} \sum_i \ln P_i + \frac{\beta}{N} \sum_i \ln |k_i|$$

In practice, since the power spectrum of a real signal is symmetric, only the positive half-space data need be considered (excluding the DC level). Thus, the data set $P_i, i = 1, 2, \ldots, N$, should be considered to include all data to the right of the DC level

Table 4.2 *Fractal dimension values.*

					$2 < k < N/2$						
D_0	1.0	1.1	1.2	1.3	1.4	1.5	1.6	1.7	1.8	1.9	2.0
D_1	1.09	1.11	1.21	1.29	1.43	1.52	1.60	1.68	1.82	1.93	1.99
					$3 < k < N/2$						
D_0	1.0	1.1	1.2	1.3	1.4	1.5	1.6	1.7	1.8	1.9	2.0
D_1	0.99	1.09	1.18	1.28	1.40	1.50	1.61	1.68	1.80	1.89	2.00
					$4 < k < N/2$						
D_0	1.0	1.1	1.2	1.3	1.4	1.5	1.6	1.7	1.8	1.9	2.0
D_1	1.02	1.10	1.19	1.28	1.36	1.51	1.63	1.74	1.78	1.91	1.97

(assuming that the power spectrum is viewed in 'optical form', where the DC level is taken to be the midpoint $1 + N/2$ of the array).

Inverse algorithm

The algorithm used to estimate the value of the spectral component β and thus $q = \beta/2$ is based on the least squares method:

Step 1 Compute the power spectrum P_i of the fractal noise.
Step 2 Extract the positive half-space data (excluding the DC level).
Step 3 Use a least squares fit to estimate $\beta = 2q$.
Step 4 Using the relationship $\beta = 5 - 2D$ provides a non-iterative formula for computing the fractal dimension from the power spectrum of a signal.

The results obtained for different ranges of k are illustrated in Table 4.2. D_0 is the orginal value of the fractal dimension and D_1 the value obtained.

4.3.2 Fractal images

Consider the case where we define a random scaling fractal image (a Mandelbrot surface) to be the solution to the equation

$$\nabla^q f(\mathbf{r}) = n(\mathbf{r}), \qquad \mathbf{r} = \hat{\mathbf{x}}x + \hat{\mathbf{y}}y \tag{4.5}$$

where $q > 0$ and

$$\nabla^q f(\mathbf{r}) \Longleftrightarrow |\mathbf{k}|^q F(\mathbf{k}), \qquad \mathbf{k} = \hat{\mathbf{x}}k_x + \hat{\mathbf{y}}k_y$$

Here k_x and k_y are the spatial frequencies in the x and y directions respectively,

$$F(\mathbf{k}) = \frac{1}{(2\pi)^2} \int_{-\infty}^{\infty} \int_{-\infty}^{\infty} f(\mathbf{r}) \exp(i\mathbf{k} \cdot \mathbf{r}) \, \mathrm{d}^2\mathbf{k}$$

and $\Longleftrightarrow$ denotes a Fourier transform pair.

Can we show that $f(\mathbf{r})$ is statistically self-affine? The general solution to Eqn 4.5 can be obtained using Fourier transforms. Thus, since in Fourier space Eqn 4.5 becomes

$$|\mathbf{k}|^q F(\mathbf{k}) = N(\mathbf{k})$$

where $N(\mathbf{k})$ is the Fourier transform of $n(\mathbf{r})$, we obtain

$$f(\mathbf{r}) = \frac{1}{(2\pi)^2} \int_{-\infty}^{\infty} \int_{-\infty}^{\infty} \frac{N(\mathbf{k})}{|\mathbf{k}|^q} \exp(i\mathbf{k} \cdot \mathbf{r})\, \mathrm{d}^2\mathbf{k}$$

or

$$f(\mathbf{r}) = \int_{-\infty}^{\infty} \int_{-\infty}^{\infty} h(\mathbf{r} - \mathbf{r}')n(\mathbf{r}')\, \mathrm{d}^2\mathbf{r}'$$

where

$$h(\mathbf{r}) = \frac{1}{(2\pi)^2} \int_{-\infty}^{\infty} \int_{-\infty}^{\infty} \frac{1}{|\mathbf{k}|^q} \exp(i\mathbf{k} \cdot \mathbf{r})\, \mathrm{d}^2\mathbf{k}$$

via the convolution theorem. However, the evaluation of $h(\mathbf{r})$ as compared with $h(x)$ (which uses some elementary results associated with the Laplace transform) is not elementary and requires the application of generalized function theory. Since the integral is of the same form for higher-order fractals, for completeness it can be shown that for n dimensions

$$\int_{-\infty}^{\infty} \int_{-\infty}^{\infty} \cdots \int_{-\infty}^{\infty} r^\beta \exp[-i(k_1 r_1 + k_2 r_2 + \cdots + k_n r_n)]\, \mathrm{d}r_1 \mathrm{d}r_2 \ldots \mathrm{d}r_n$$
$$= \frac{(\frac{1}{2}\beta + \frac{1}{2}n - 1)!}{(-\frac{1}{2}\beta - 1)!} 2^{\beta+n} \pi^{n/2} k^{-\beta-n},$$

for $\beta \neq 2m$ or $-n - 2m$ and $m = 0, 1, 2, \ldots$; $r = \sqrt{r_1^2 + r_2^2 + \cdots + r_n^2}$ and $k = \sqrt{k_1^2 + k_2^2 + \cdots + k_n^2}$.

Hence, for the two-dimensional case,

$$\int_{-\infty}^{\infty} \int_{-\infty}^{\infty} |\mathbf{r}|^{q-2} \exp(i\mathbf{k} \cdot \mathbf{r})\, \mathrm{d}^2\mathbf{r} = \frac{\alpha}{|\mathbf{k}|^q}$$

where (using Gamma function notation)

$$\alpha(q) = \frac{2^q \pi \Gamma(\frac{1}{2}q)}{\Gamma(1 - \frac{1}{2}q)}$$

and therefore the desired result is

$$h(\mathbf{r}) = \frac{1}{\alpha(q)}|\mathbf{r}|^{2-q}$$

The fractal image $f(\mathbf{r})$ can then be expressed in the form

$$f(\mathbf{r}) = \frac{1}{\alpha(q)} \int_{-\infty}^{\infty}\int_{-\infty}^{\infty} \frac{n(\mathbf{r}')}{|\mathbf{r}-\mathbf{r}'|^{2-q}} \mathrm{d}^2\mathbf{r}'$$

This equation has the characteristic scaling property:

$$f'(\mathbf{r}) = \frac{1}{\alpha(q)} \int_{-\infty}^{\infty}\int_{-\infty}^{\infty} \frac{n(\lambda\mathbf{r}')}{|\mathbf{r}-\mathbf{r}'|^{2-q}} \mathrm{d}^2\mathbf{r}'$$

$$= \frac{1}{\lambda^q\alpha(q)} \int_{-\infty}^{\infty}\int_{-\infty}^{\infty} \frac{n(\mathbf{y})}{|\lambda\mathbf{r}-\mathbf{y}|^{2-q}} \mathrm{d}^2\mathbf{y} = \frac{1}{\lambda^q} f(\lambda\mathbf{r})$$

where $\mathbf{y} = \lambda\mathbf{r}'$. Thus

$$\Pr[f_\lambda(\mathbf{r})] = \frac{1}{\lambda^q}\Pr[f(\lambda\mathbf{r})]$$

which describes a statistically self-affine (Mandelbrot) surface or fractal image. Note that the n-dimensional Fourier transform of r^{q-n} is proportional to k^{-q} and so (ignoring scaling) an n-dimensional random scaling fractal (RSF) can be written in terms of the n-dimensional convolution of n-dimensional white noise with r^{n-q}. That is, an RSF signal is the convolution of $n(x)$ with x^{1-q}; an RSF surface (or image) is the convolution of $n(\mathbf{r})$ with r^{2-q}; an RSF volume is the convolution of $n(\mathbf{r})$ with r^{3-q} and so on.

Developing mathematical models to simulate and analyse noise has an important role in digital signal and image processing. Computer generated noise is routinely used to test the robustness of different types of algorithm (e.g. algorithms whose principal goal is to extract information from noise); it is also used for data encryption and is even used to enhance or amplify signals through 'stochastic resonance'. Accurate statistical models for noise (e.g. the probability distribution function or the characteristic function) are particularly important in image restoration using Bayesian estimation [36], in maximum-entropy methods for signal and image reconstruction [37] and in the image segmentation of coherent images in which 'speckle' (arguably a special type of noise, i.e. coherent Gaussian noise) is a prominent feature [38]. The noise characteristics of a given imaging system often dictate the type of filters that are used to process and analyse the data. Noise simulation is also important in the synthesis of images used in computer graphics and computer animation systems, in which fractal noise has a special place (e.g. [1] and [39]).

The application of fractal geometry for modelling naturally occurring signals and images is well known. This is due to the fact that the 'statistics' and spectral characteristics of RSFs are consistent with many objects found in nature, a characteristic that is expressed in the term 'statistical self-affinity'. This term refers to random processes whose statistics are scale invariant. An RSF signal is one whose probability density function (PDF) remains the same irrespective of the scale over which the signal is sampled. Thus, as we zoom into an RSF signal, although the time signature changes the PDF of the signal remains the same (a scaled-down version of the original) – a concept that is aptly expressed in the Chinese proverb 'In every way one can see the shape of the sea.'

The many signals found in nature that are statistically self-affine include a wide range of noise sources, including background cosmic radiation (at most frequencies) for example. In addition, certain speech signals (representative of fricatives) exhibit the characteristics of RSFs, as do other signals such as financial time series, seismic signals and so on. The incredible range of vastly different systems that exhibit random fractal behaviour is leading more and more researchers to consider statistical self-affinity to be a universal law, a law that is particularly evident in systems that are undergoing a phase transition (see e.g. [40, 41]).

In a stable state, the behaviour of the elements of which a system is composed depends primarily on their local neighbours, and so the statistics of the system are not self-affine. In a critical state, the elements become connected over long ranges, propagating 'order' throughout the system in the sense that the statistical characteristics of the system are self-affine with system-wide correlations. This is more to do with the connectivity of the elements than the elements themselves [42]. (Critical states can of course be stable in the dynamical sense.) Moreover, critical states appear to be governed by the universal power law

$$System \propto \frac{1}{Size^q}$$

where q is a non-integer value and the term *System* is a generic term representative of some definable parameter that can be measured experimentally over different scales of a certain *Size*. This power law is the principal signature that the system is behaving in a statistically self-affine way. There is a wide variety of examples that demonstrate this power law. For example, the growth rate of companies tends to diminishes with size, irrespective of the type of business being conducted: typical US companies are characterized by $q \in [0.7, 0.9]$. The frequency of the creation and extinction of species (as revealed through the growing number of fossil records) is starting to indicate that the pattern of fitness for survival is statistically self-affine. This result can be simulated by relatively simple iteration-function systems such as in the Bak–Sneppen model, which yields self-affine distributions for the fitness of different species over time [42]. The distribution of base pairs in DNA is statistically self-affine, i.e. the frequency of occurrence of adenine–thymine and cytosine–guanine in a DNA molecule is the same at different scales. DNA is, in effect, a self-affine bit stream.

Conventional RSF models are based on *stationary* processes, in which the 'statistics' of the signal are invariant in time. However, many signals exhibit non-stationary behaviour. In addition, many signals exhibit episodes that are rare but extreme (sudden changes in amplitude and/or frequency), events that are statistically inconsistent with the 'normal' behaviour of the signal. These episodes include so-called Lévy-type flights in cases when the statistics of the signal conform to that of a Lévy-type distribution. Lévy statistics are the basis for one of the power-law-based stochastic models that are currently being used to investigate the 'strange kinetics' of systems undergoing phase transitions, including hot plasmas, superfluids, superconducting materials and economic systems (see e.g. [43, 44]).

4.3.3 Stochastic modelling

There are two principal criteria used to define the characteristics of a stochastic field:

- the PDF or equivalently the characteristic function (i.e. the Fourier transform of the PDF)
- the power spectral density function (PSDF)

The PSDF is the function that describes the envelope or shape of the power spectrum of the field and is related to the autocorrelation function of a signal through the autocorrelation theorem. In this sense the PSDF measures the time correlations of a signal. For example, zero-mean white Gaussian noise is a stochastic field characterized by a PSDF that is effectively constant over all frequencies and has a PDF with a Gaussian profile whose mean is zero.

Stochastic fields can of course be characterized using transforms other than the Fourier transform (from which the PSDF is obtained), such as the Wigner transform [45]. Wavelets are also being used effectively to analyse and process noise in signals and images (e.g. [46] and references therein). However, the conventional PDF–PSDF approach serves many purposes; providing a detailed comparison with other techniques of stochastic analysis is beyond the scope of this book.

There are two conventional approaches to simulating a stochastic field. The first is based on predicting the PDF (or the characteristic function) theoretically, if possible. A pseudo-random-number generator is then designed whose output provides a discrete stochastic field characteristic of the predicted PDF. For example, a Gaussian pseudo-random-number generator can be designed using the box–Müller transformation operating on uniform deviates [47]. The second approach is based on considering the PSDF of a signal, which, like the PDF, is ideally derived theoretically. The stochastic field is then simulated, typically by filtering white Gaussian noise.

Many stochastic fields observed in nature have two fundamental properties:

- the PSDF is determined by irrational-number power laws such as $1/|\omega|^q$, which describes a certain type of noise; here ω is the angular frequency and q is an irrational number;
- the field is statistically self-affine.

4.3.4 A new non-stationary fractional dynamic model

A common theme associated with the fractional dynamic models discussed in the previous subsection is that they describe stationary random processes in terms of solutions to a PDF. In this subsection, we postulate a partial differential equation whose characteristics incorporate behaviour that describes non-stationary fractal walks and Lévy-type flights in terms of a solution to the stochastic field itself (and not its PDF). Within the context of the solution derived in this chapter, these so-called Lévy-type flights are actually the result of randomly introducing Brownian noise over a short period of time into an otherwise non-stationary fractal signal. In this sense, they have nothing to do with Lévy flights as such, but produce results that may be considered to be analogous to them. These effects are called 'Brownian transients' and have longer time correlations than fractal noise.

Suppose we consider an inhomogeneous fractal diffusion equation of the form

$$\left[\frac{\partial^2}{\partial x^2} - \tau\frac{\partial^q}{\partial t^q}\right]u(x,t) = -F(x,t), \qquad 0 < q \leq 1$$

where τ is a constant, F is a stochastic source term having some PDF and u is the stochastic field whose solution we require. When $q = 1$ we obtain the diffusion equation but, in general, a solution to this equation will provide stationary temporal fractal walks – random walks of fractal time. One way of introducing a (first-order) non-stationary process is to let $F(x,t)$ be a functional, i.e. to consider a source term of the form $F[x, t, \alpha(x), \beta(t)]$, where α and β are arbitrary functions. For example, suppose that we write F in separable form $F(x,t) = f(x)n(t)$, where $n(t)$ is a random variable of time with the usual PDF given by

$$\Pr[n(t)] = \frac{1}{\sigma\sqrt{2\pi}}\exp\left[\frac{-(\mu-n)^2}{2\sigma^2}\right], \qquad -\infty < n < \infty$$

Here μ and σ are the mean and standard deviations respectively. By letting μ and/or σ be functions of t, we introduce time variations into the mean and/or standard deviation respectively. In this case, varying the mean will cause the range of $n(t)$ to change with time and varying the standard deviation will change the variance of $n(t)$ with time. Note that the form of the distribution of the field remains the same – it is a time-varying Gaussian field. A more general statement of a non-stationary process is one in which the distribution itself changes with time.

Another way of introducing non-stationarity is through q, by letting it become a function of time t. Suppose that, in addition to this, we extend the range of q to include the values 0 and 2 so that $0 \leq q \leq 2$. This idea immediately leads us to an interesting consequence because, with q in this range, we could choose $q = 1$ to get the diffusion equation but could instead choose $q = 2$ to obtain an entirely different equation, namely, the wave equation. Choosing (quite arbitrarily) q to be in this range leads to control over the basic physical characteristics of the equation, so that we can define a statistical mode when $q = 0$, a diffusive mode when $q = 1$ and a propagative mode when $q = 2$.

In this case, non-stationarity is introduced through the use of a time-varying fractional derivative whose values can change the physical meaning of the equation. Since the range of q has been chosen arbitrarily, we generalize further and consider the equation

$$\left[\frac{\partial^2}{\partial x^2} - \tau^{q(t)}\frac{\partial^{q(t)}}{\partial t^{q(t)}}\right]u(x,t) = -F(x,t), \qquad -\infty < q(t) < \infty, \quad \forall\, t \tag{4.6}$$

As before, when $q = 0\ \forall\, t$, the time-dependent behaviour is determined by the source function alone; when $q = 1\ \forall\, t$, u describes a diffusive process where τ is the 'diffusivity' (the inverse of the coefficient of diffusion); when $q = 2$ we have a propagative process where τ is the 'slowness' (the inverse of the wave speed). The latter process should be expected to propagate information more rapidly than a diffusive process leading to transients or 'flights' of some type. We shall refer to the parameter q as the *Fourier dimension*, which, for a continuous fractal signal, is related to the conventional definition of the fractal dimension D (i.e. the similarity, Minkowksi or box-counting) by [17]

$$q \le D + \tfrac{1}{2}, \qquad 1 < D < 2 \tag{4.7}$$

Choosing the Fourier dimension $q(t)$

Since the Fourier dimension $q(t)$ drives the non-stationary behaviour of u, the way in which we model $q(t)$ is crucial. It is arguable that the changes in the statistical characteristics of u that lead to its non-stationary behaviour should in themselves be random. Thus, suppose that we let $q(t)$ at a time t be chosen randomly, with a randomness that is determined by some PDF. In this case, the non-stationary characteristics of u will be determined by the PDF (and associated parameters) alone. Since q is a dimension, we can consider our model to be based on the 'statistics of dimension'.

There is a variety of PDFs that could be applied (including a Lévy distribution); the choice of PDF will in turn affect the range of q. By varying the exact nature of the distribution considered, we can 'drive' the non-stationary behaviour of u in different ways. For example, suppose we consider a system that is assumed to be primarily diffusive; then a 'normal' PDF, of the type

$$\Pr[q(t)] = \frac{1}{\sigma\sqrt{2\pi}} \exp\left[\frac{-(q-1)^2}{2\sigma^2}\right], \qquad -\infty < q < \infty$$

will ensure that u is entirely diffusive when $\sigma \to 0$. However, as σ is increased in value, the likelihood of $q = 2$ (and $q = 0$) becomes larger. In other words, the standard deviation provides control over the likelihood that the process will become propagative. If, for example, we consider a gamma distribution, given by

$$\Pr[q(t)] = \begin{cases} \dfrac{1}{\beta^\alpha}\dfrac{1}{\Gamma(\alpha)} q^{\alpha-1}\exp(-q/\beta), & q > 0 \\ 0, & q \le 0 \end{cases}$$

where $\alpha > 0$ and $\beta > 0$, then q lies in the positive half-space alone, with mean and variance given by

$$\mu = \alpha\beta \qquad \text{and} \qquad \sigma^2 = \alpha\beta^2$$

respectively. Probability density functions could also be considered that are of compact support, such as the beta distribution, given by

$$\Pr[q(t)] = \begin{cases} \dfrac{\Gamma(\alpha+\beta)}{\Gamma(\alpha)\Gamma(\beta)} q^{\alpha-1}(1-q)^{\beta-1}, & 0 < q < 1 \\ 0, & \text{otherwise} \end{cases}$$

where α and β are positive constants. Here, the mean and variance are

$$\mu = \frac{\alpha}{\alpha+\beta} \qquad \text{and} \qquad \sigma^2 = \frac{\alpha\beta}{(\alpha+\beta)^2(\alpha+\beta+1)}$$

respectively, and for $\alpha > 1$ and $\beta > 1$ there is a unique mode at the value

$$x_{\text{mode}} = \frac{\alpha - 1}{\alpha + \beta + 2}$$

Irrespective of the type of distribution that is considered, Eqn 4.6 poses a fundamental problem, which is how to define and work with the term

$$\frac{\partial^{q(t)}}{\partial t^{q(t)}} u(x, t)$$

Given the result, for constant q,

$$\frac{\partial^q}{\partial t^q} u(x, t) = \frac{1}{2\pi} \int_{-\infty}^{\infty} (i\omega)^q U(x, \omega) \exp(i\omega t) \, d\omega, \qquad -\infty < q < \infty$$

we might generalize it as follows:

$$\frac{\partial^{q(t)}}{\partial t^{q(t)}} u(x, t) = \frac{1}{2\pi} \int_{-\infty}^{\infty} (i\omega)^{q(t)} U(x, \omega) \exp(i\omega t) \, d\omega$$

However, if we consider the case where the Fourier dimension is a relatively slowly varying function of t, then we can legitimately consider $q(t)$ to be composed of a sequence of different states $q_i = q(t_i)$. This allows us to develop a stationary solution for a fixed q over a fixed period of time. Non-stationary behaviour can then be introduced by using the same solution for different values or 'quanta' q_i over fixed (or varying) periods of time and concatenating the solutions for all q_i. Quantizing $q(t)$ in this way also allows us to define a process that we call 'fractal modulation', whereby $q(t)$ is assigned two states, q_1 and q_2 where $q_1 \neq q_2$. Typically the probability of obtaining q_1 or q_2 is considered to be 0.5 so that no weighting is attributed to either state. By letting these states correspond to 0 and 1 in a bit stream, we can consider the application of fractal modulation to a digital communications systems.

4.3.5 Green's function solution

We shall consider a Green's function solution to Eqn 4.6 for constant q when $F(x,t) = f(x)n(t)$, where $f(x)$ and $n(t)$ are both stochastic functions and where $n(t)$ is taken to be white Gaussian noise. Applying a separation of variables here is not strictly necessary. However, it yields a solution in which the terms affecting the temporal behaviour of $u(x,t)$ are clearly identifiable. Thus, we require a general solution to the equation

$$\left(\frac{\partial^2}{\partial x^2} - \tau^q \frac{\partial^q}{\partial t^q}\right) u(x,t) = -f(x)n(t)$$

Let

$$u(x,t) = \frac{1}{2\pi}\int_{-\infty}^{\infty} U(x,\omega)\exp(i\omega t)\,\mathrm{d}\omega, \qquad n(t) = \frac{1}{2\pi}\int_{-\infty}^{\infty} N(\omega)\exp(i\omega t)\,\mathrm{d}\omega$$

Then, using the result

$$\frac{\partial^q}{\partial t^q}u(x,t) = \frac{1}{2\pi}\int_{-\infty}^{\infty} U(x,\omega)(i\omega)^q \exp(i\omega t)\,\mathrm{d}\omega$$

the fractional PDE transforms to

$$\left(\frac{\partial^2}{\partial x^2} + \Omega_q^2\right) U(x,\omega) = -f(x)N(\omega)$$

where we shall take

$$\Omega_q = i(i\omega\tau)^{q/2}$$

and ignore the case $\Omega_q = -i(i\omega\tau)^{q/2}$. Defining the Green's function g to be the solution of

$$\left(\frac{\partial^2}{\partial x^2} + \Omega_q^2\right) g(x|x_0,\omega) = -\delta(x - x_0)N(\omega)$$

where δ is the delta function, we obtain the following solution:

$$U(x_0,\omega) = N(\omega)\int_{-\infty}^{\infty} g(x|x_0,k)f(x)\,\mathrm{d}x \tag{4.8}$$

with

$$g(X,\omega) = \frac{1}{2\pi}\int_{-\infty}^{\infty} \frac{\exp(iuX)}{(u+\Omega_q)(u-\Omega_q)}\,du, \qquad X = |x - x_0|$$

The contour integral

$$\oint_C \frac{\exp(izX)}{(z+\Omega_q)(z-\Omega_q)}\,\mathrm{d}z$$

has complex poles at $z = \pm\Omega_q$ that are q-dependent (varying from $\pm i$ when $q = 0$ through to $\pm i(i\omega\tau)^{1/2}$ when $q = 1$ and on to $\mp\omega\tau$ when $q = 2$, for example). For any value of q, we can compute this contour integral using the residue theorem. The contour must be chosen, of course, in such a way that it runs along the real axis in order to evaluate g. By choosing to evaluate the integral for a q-dependent pole in the z plane for $-\infty < x < \infty$ and $0 \leq iy < \infty$ where $z = x + iy$ we obtain (using a semicircular contour C)

$$g(x|x_0, k) = \frac{i}{2\Omega_q} \exp(i\Omega_q |x - x_0|) \tag{4.9}$$

under the assumption that Ω_q is finite. This result reduces to the conventional solutions for the cases $q = 1$ (the diffusion equation) and $q = 2$ (the wave equation), as shall now be shown.

Wave equation solution

When $q = 2$, Eqn 4.8 provides a solution for the outgoing Green's function [48]. Thus, with $\Omega_2 = -\omega\tau$, we have

$$U(x_0, \omega) = \frac{N(\omega)}{2i\omega\tau} \int_{-\infty}^{\infty} \exp(-i\omega\tau|x - x_0|) f(x)\,\mathrm{d}x$$

and Fourier-inverting we get

$$\begin{aligned} u(x_0, t) &= \frac{1}{2\tau} \int_{-\infty}^{\infty} \mathrm{d}x\, f(x) \frac{1}{2\pi} \int_{-\infty}^{\infty} \frac{N(\omega)}{i\omega} \exp(-i\omega\tau|x - x_0|) \exp(i\omega t)\,\mathrm{d}\omega \\ &= \frac{1}{2\tau} \int_{-\infty}^{\infty} \mathrm{d}x\, f(x) \int_{-\infty}^{t} n(t - \tau|x - x_0|)\,\mathrm{d}t \end{aligned}$$

which describes the propagation of a wave travelling at velocity $1/\tau$ subject to variations in space and time defined by $f(x)$ and $n(t)$ respectively. For example, when f and n are both delta functions,

$$u(x_0, t) = \frac{1}{2\tau} H(t - \tau|x - x_0|)$$

where H is the Heaviside step function defined by

$$H(y) = \begin{cases} 1, & y > 0 \\ 0, & y < 0 \end{cases}$$

This is a d'Alembertian-type solution to the wave equation, in which the wavefront occurs at $t = \tau|x - x_0|$ in the causal case.

Diffusion equation solution

When $q = 1$ and $\Omega_1 = i\sqrt{i\omega\tau}$,

$$u(x_0, t) = \frac{1}{2}\int_{-\infty}^{\infty} \mathrm{d}x\, f(x)\frac{1}{2\pi}\int_{-\infty}^{\infty} \frac{\exp(-\sqrt{i\omega\tau}|x - x_0|)}{\sqrt{i\omega\tau}} N(\omega)\exp(i\omega t)\,\mathrm{d}\omega$$

For $p = i\omega$, we can write this result in terms of a Bromwich integral (i.e. an inverse Laplace transform) and use the convolution theorem for Laplace transforms to give [49]

$$\int_{c-i\infty}^{c+i\infty} \frac{\exp(-a\sqrt{p})}{\sqrt{p}} \exp(pt)\,\mathrm{d}p = \frac{1}{\sqrt{\pi t}} \exp\left(\frac{-a^2}{4t}\right)$$

We obtain

$$u(x_0, t) = \frac{1}{2\sqrt{\tau}}\int_{-\infty}^{\infty} \mathrm{d}x\, f(x)\int_{0}^{t} \frac{1}{\sqrt{\pi t_0}} \exp\left[\frac{-\tau(x_0 - x)^2}{\sqrt{4t_0}}\right] n(t - t_0)\,\mathrm{d}t_0$$

Thus, if for example we consider the case where n is a delta function, the result reduces to

$$u(x_0, t) = \frac{1}{2\sqrt{\pi\tau t}}\int_{-\infty}^{\infty} f(x)H(t)\exp\left[\frac{-\tau(x_0 - x)^2}{4t}\right] \mathrm{d}x, \qquad t \to \infty$$

which describes classical diffusion in terms of the convolution of an initial source $f(x)$ (introduced at time $t = 0$) with a Gaussian function.

General series solution

The evaluation of $u(x_0, t)$ from Eqn 4.8 via direct Fourier inversion for arbitrary values of q is not possible, owing to the irrational nature of the exponential function $\exp(i\Omega_q|x - x_0|)$ with respect to ω. To obtain a general solution, we use the series representation of the exponential function and write Eqn 4.8 in the form

$$U(x_0, \omega) = \frac{iM_0N(\omega)}{2\Omega_q}\left[1 + \sum_{m=1}^{\infty} \frac{(i\Omega_q)^m}{m!}\frac{M_m(x_0)}{M_0}\right]$$

where

$$M_m(x_0) = \int_{-\infty}^{\infty} f(x)|x - x_0|^m dx$$

Then we Fourier-invert term by term to develop a series solution. This requires us to consider three distinct cases.

1. When $q = 0$, evaluation of $u(x_0, t)$ is trivial, since from Eqn 4.8

$$U(x_0, \omega) = \frac{M(x_0)}{2} N(\omega) \qquad \text{or} \qquad u(x_0, t) = \frac{M(x_0)}{2} n(t) \tag{4.10}$$

where

$$M(x_0) = \int_{-\infty}^{\infty} \exp(-|x - x_0|) f(x) \, dx$$

2. When $q > 0$, Fourier-inverting, the first term in this series becomes

$$\frac{1}{2\pi} \int_{-\infty}^{\infty} \frac{i N(\omega) M_0}{2\Omega_q} \exp(i\omega t) \, d\omega = \frac{M_0}{2\tau^{q/2}} \frac{1}{2\pi} \int_{-\infty}^{\infty} \frac{N(\omega)}{(i\omega)^{q/2}} \exp(i\omega t) \, d\omega$$

$$= \frac{M_0}{2\tau^{q/2}} \frac{1}{\Gamma(q/2)} \int_0^t \frac{n(\xi)}{(t - \xi)^{1-(q/2)}} \, d\xi, \qquad \text{Re}\, q > 0$$

The second term is

$$-\frac{M_1}{2} \frac{1}{2\pi} \int_{-\infty}^{\infty} N(\omega) \exp(i\omega t) \, d\omega = -\frac{M_1}{2} n(t)$$

The third term is

$$-\frac{i M_2}{2 \times 2!} \frac{1}{2\pi} \int_{-\infty}^{\infty} N(\omega) i (i\omega\tau)^{q/2} \exp(i\omega t) \, d\omega = \frac{M_2 \tau^{q/2}}{2 \times 2!} \frac{d^{q/2}}{dt^{q/2}} n(t)$$

and the fourth and fifth terms become

$$\frac{M_3}{2 \times 3!} \frac{1}{2\pi} \int_{-\infty}^{\infty} N(\omega) i^2 (i\omega\tau)^q \exp(i\omega t) \, d\omega = -\frac{M_3 \tau^q}{2 \times 3!} \frac{d^q}{dt^q} n(t)$$

and

$$i \frac{M_4}{2 \times 4!} \frac{1}{2\pi} \int_{-\infty}^{\infty} N(\omega) i^3 (i\omega\tau)^{3q/2} \exp(i\omega t) \, d\omega = \frac{M_4 \tau^{3q/2}}{2 \times 4!} \frac{d^{3q/2}}{dt^{3q/2}} n(t)$$

respectively, with similar results for all other terms. Thus, through induction, we can write $u(x_0, t)$ as a series of the form

$$u(x_0, t) = \frac{M_0(x_0)}{2\tau^{q/2}} \frac{1}{\Gamma(q/2)} \int_0^t \frac{n(\xi)}{(t - \xi)^{1-(q/2)}} \, d\xi - \frac{M_1(x_0)}{2} n(t)$$
$$+ \frac{1}{2} \sum_{k=1}^{\infty} \frac{(-1)^{k+1}}{(k+1)!} M_{k+1}(x_0) \tau^{kq/2} \frac{d^{kq/2}}{dt^{kq/2}} n(t) \tag{4.11}$$

Observe that the first term involves a fractional integral, the second term consists of the source function $n(t)$ alone (apart from a scaling factor) and the third term is an infinite series composed of fractional differentials of increasing order $kq/2$. Note also that the first term is scaled by a factor involving $\tau^{-q/2}$ whereas the third term is scaled by a factor that includes $\tau^{kq/2}$.

3. When $q < 0$, the first term becomes

$$\frac{1}{2\pi}\int_{-\infty}^{\infty}\frac{iN(\omega)M_0}{2\Omega_q}\exp(i\omega t)\,\mathrm{d}\omega = \frac{M_0}{2}\tau^{q/2}\frac{1}{2\pi}\int_{-\infty}^{\infty}N(\omega)(i\omega)^{q/2}\exp(i\omega t)\,\mathrm{d}\omega$$
$$= \frac{M_0}{2}\tau^{q/2}\frac{\mathrm{d}^{q/2}}{\mathrm{d}t^{q/2}}n(t)$$

The second term remains the same and the third term is

$$-\frac{iM_2}{2\times 2!}\frac{1}{2\pi}\int_{-\infty}^{\infty}\frac{N(\omega)i}{(i\omega\tau)^{q/2}}\exp(i\omega t)\,\mathrm{d}\omega = \frac{M_2}{2\times 2!}\frac{1}{\tau^{q/2}}\frac{1}{\Gamma(q/2)}\int_0^t\frac{n(\xi)}{(t-\xi)^{1-(q/2)}}\,\mathrm{d}\xi$$

Evaluating the other terms by induction we obtain

$$u(x_0,t) = \frac{M_0(x_0)\tau^{q/2}}{2}\frac{\mathrm{d}^{q/2}}{\mathrm{d}t^{q/2}}n(t) - \frac{M_1(x_0)}{2}n(t)$$
$$+\frac{1}{2}\sum_{k=1}^{\infty}\frac{(-1)^{k+1}}{(k+1)!}\frac{M_{k+1}(x_0)}{\tau^{kq/2}}\frac{1}{\Gamma(kq/2)}\int_0^t\frac{n(\xi)}{(t-\xi)^{1-(kq/2)}}\,\mathrm{d}\xi \tag{4.12}$$

where $q \equiv |q|, q < 0$. Here, the solution is composed of three terms, a fractional differential, the source term and an infinite series of fractional integrals of order $kq/2$. Thus, the roles of fractional differentiation and fractional integration are reversed as q changes from being greater than zero to less than zero.

Asymptotic forms for $f(x) = \delta(x)$

Eqns. 4.10, 4.11 and 4.12 warrant further theoretical investigation, which is beyond the scope of this book. Instead, we consider a special case in which the source function $f(x)$ is an impulse, so that

$$M_m(x_0) = \int_{-\infty}^{\infty}\delta(x)|x-x_0|^m dx = |x_0|^m$$

This result immediately suggests a study of the asymptotic solution

$$u(t) = \lim_{x_0 \to 0} u(x_0, t) = \begin{cases} \dfrac{1}{2\tau^{q/2}} \dfrac{1}{\Gamma(q/2)} \displaystyle\int_0^x \dfrac{n(\xi)}{(t-\xi)^{1-(q/2)}} \mathrm{d}\xi, & q > 0 \\ \dfrac{n(t)}{2}, & q = 0 \\ \dfrac{\tau^{q/2}}{2} \dfrac{\mathrm{d}^{q/2}}{\mathrm{d}t^{q/2}} n(t), & q < 0 \end{cases}$$

The solution for the time variations of the stochastic field u for $q > 0$ is then given by a fractional integral alone and for $q < 0$ by a fractional differential alone. In particular, for $q > 0$ we see that the solution is based on a causal convolution. Thus in t-space

$$u(t) = \frac{1}{2\tau^{q/2}\Gamma(q)} \frac{1}{t^{1-q/2}} \otimes n(t), \qquad q > 0$$

where $\otimes$ denotes (causal) convolution, and in ω-space

$$U(\omega) = \frac{N(\omega)}{2\tau^{q/2}(i\omega)^{q/2}} \tag{4.13}$$

This result is the conventional fractal-noise model where, for a fractal signal, q is related to the fractal dimension. Table 4.3 quantifies the results for different values of q, with conventional name associations for the noise. Note that u has the following fundamental property:

$$\lambda^q \Pr[u_\lambda(t)] = \Pr[u(\lambda t)]$$

where

$$u_\lambda(t) = \frac{1}{2\tau^{q/2}\Gamma(q)} \frac{1}{t^{1-q/2}} \otimes n(\lambda t), \qquad \lambda > 0$$

This property describes the statistical self-affinity of u. Thus, the asymptotic solution considered here yields a result that describes an RSF signal characterized by a PSDF of the form $1/|\omega|^q$, which is a measure of the time correlations in the signal.

Other asymptotic forms

Another interesting asymptotic form of Eqn 4.11 is

$$u(x_0, t) = \frac{M_0(x_0)}{2\tau^{q/2}} \frac{1}{\Gamma(q/2)} \int_0^t \frac{n(\xi)}{(t-\xi)^{1-(q/2)}} \mathrm{d}\xi - \frac{M_1(x_0)}{2} n(t), \qquad \tau \to 0 \tag{4.14}$$

Table 4.3 *Noise characteristics for different values of q (note that $\Gamma(1/2)=\sqrt{\pi}$ and $\Gamma(1)=1$)*

q-value	t-space	ω-space (PSDF)	Name
$q=0$	$\frac{1}{2}n(t)$	$\frac{1}{4}$	white noise
$q=1$	$\frac{1}{2\sqrt{\tau}\Gamma(1/2)}\frac{1}{\sqrt{t}}\otimes n(t)$	$\frac{1}{4\tau\lvert\omega\rvert}$	pink noise
$q=2$	$\frac{1}{2\tau\Gamma(1)}\int_0^t n(t)\,\mathrm{d}t$	$\frac{1}{4\tau^2\omega^2}$	brown noise
$q>2$	$\frac{1}{2\tau^{q/2}\Gamma(q/2)}t^{(q/2)-1}\otimes n(t)$	$\frac{1}{4\tau^q\lvert\omega\rvert^q}$	black noise

Here, the solution is the sum of fractal noise and white noise. By relaxing the condition $\tau \to 0$ we can consider the approximation

$$u(x_0,t) \simeq \frac{M_0(x_0)}{2\tau^{q/2}}\frac{1}{\Gamma(q/2)}\int_0^t \frac{n(\xi)}{(t-\xi)^{1-(q/2)}}\,\mathrm{d}\xi - \frac{M_1(x_0)}{2}n(t) + \frac{M_2(x_0)}{2\times 2!}\tau^{q/2}\frac{\mathrm{d}^{q/2}}{\mathrm{d}t^{q/2}}n(t), \qquad \tau \ll 1$$

in which the solution is expressed in terms of the sum of fractal noise, white noise and the fractional differentiation of white noise.

REFERENCES

[1] H. O. Peitgen and D. Saupe. *The Science of Fractal Images*. Springer-Verlag, 1988.

[2] N. Yokoya, F. Yamamoto and N. Funakubo. Fractal based analysis and interpolation of 3D natural surface shapes. *Computer Vision, Graphics and Image Processing*, **46**: 284–302, 1989.

[3] B. B. Mandelbrot. *The Fractal Geometry of Nature*. Freeman, 1983.

[4] L. F. Richardson, ed. *The Problem of Contiguity: An Appendix to Statistics of Deadly Quarrels*, vol. 6. General Systems Yearbook, 1961.

[5] A. Rosenfeld and E. B. Troy. Visual texture analysis. Technical Report TR-116, University of Maryland, 1970.

[6] K. C. Hayes *et al.* Texture coarseness: further experiments. *IEEE Trans. Systems, Man and Cybernetics*, **4**: 467–72, 1974.

[7] A. Rosenfeld and A. C. Kak. *Digital Picture Processing*. New York: Academic, 1976.

[8] R. Bajcsy. Computer description of textured surfaces. In *Proc. Int. Conf. on Artificial Intelligence*, pp. 572–9, 1973.

[9] R. Bajcsy. Computer identification of textured visual scenes. Technical Report AIM-180, Artificial Intelligence Laboratory, Stanford University, 1972.

[10] R. M. Haralick. Statistical image texture analysis. In T. Y. Young and K.-S. Fu, eds, *Handbook of Pattern Recognition and Image Processing*. Academic Press, 1986.

[11] M. Reuff. Scale space filtering and the scaling regions of fractals. In J. C. Simon, ed., *From Pixels to Features*, pp. 49–60, North Holland, 1989.

[12] P. Maragos and F. K. Sun. Measuring fractal dimension: morphological estimates and iterative estimation, In *Visual Communications and Image Processing IV, Proc. SPIE 89*, vol. 11, pp. 416–30, 1989.

[13] P. P. Ohanian and R. C. Dubes. Performance evaluation for four classes of textural features. *Pattern Recognition*, **25**(8): 819–33, 1992.

[14] P. Kube and A. Pentland. On the imaging of fractal surfaces. *IEEE Trans. Pattern Analysis and Machine Intelligence*, **10**(5): 704–7, 1988.

[15] K. B. Oldham and J. Spanier. *The Fractional Calculus*, vol. 3 of *Mathematics in Science and Engineering*. Academic Press, 1974.

[16] S. G. Samko, AA. Kilbas and O. I. Marichev. *Fractional Integrals and Derivatives: Theory and Applications*. Gordon and Breach, 1993.

[17] A. Evans. Fourier dimension and fractal dimension. *J. Chaos, Solitons and Fractals*, **9**(12): 1977–82, 1998.

[18] M. Goodchild. Fractals and the accuracy of geographical measures. *Mathematical Geology*, **12**(2): 85–98, 1980.

[19] M. Shelberg. The development of a curve and surface algorithm to measure fractal dimensions. Master's thesis, Ohio State University, 1982.

[20] K. Clarke and D. Schweizer. Measuring the fractal dimension of natural surfaces using a robust fractal estimator. *Cartography and Geographic Information Systems*, **18**(1): 27–47, 1991.

[21] N. Lam. and L. De Cola. *Fractals in Geography*. Prentice-Hall, 1993.

[22] R. Voss. Random fractals: characterisation and measurement in scaling phenomena in disordered systems. In R. Pynn and A. Skjeltorps, eds., *Scaling Phenomena in Disordered Systems*. Plenum, 1986.

[23] N. Sarkar and B. Chaudhuri. An efficient approach to estimate fractal dimension of textured images. *Pattern Recognition*, **25**(9): 1035–41, 1992.

[24] N. Sarker, B. Chaudhuri, and P. Kundu. Improved fractal geometry based texture segmentation technique. *Proc. IEEE*, **140**(5): 233–41, 1993.

[25] J. Keller and S. Chen. Texture description and segmentation through fractal geometry. *Computer Vision, Graphics and Image Processing*, **45**: 150–66, 1989.

[26] K. Clarke. Scale based simulation of topographic relief. *The American Cartographer*, **15**(2):173–81, 1988.

[27] M. Shelberg, N. Lam and H. Moellering. Measuring the fractal dimensions of surfaces. In *Proc. 6th Int. Symp. on Automated Cartography, Ottawa*, pp. 319–28, 1993.

[28] N. Lam. Description and measurement of landsat tm images using fractals. *Photogrammetric Engineering and Remote Sensing*, **56**(2): 187–95, 1990.

[29] S. Jaggi, D. Quattrochi and N. Lam. Implementation and operation of three fractal measurement algorithms for analysis of remote sensing data. *Computers and Geosciences*, **19**(6): 745–67, 1993.

[30] L. DeCola. Fractal analysis of a classified landsat scene. *Photogrammetric Engineering and Remote Sensing*, **55**(5): 601–10, 1989.

[31] T. Peli, V. Tom and V. Lee. Multi-scale fractal and correlation signatures for image screening and natural clutter suppression. In *Proc. SPIE 89, Visual Communication and Image Processing IV*, pp. 402–15, 1989.

[32] A. M. Vepsäläinen and J. Ma. Estimating the fractal and correlation dimension from 2D and 3D images. In *Proc. SPIE 89, Visual Communication and Image Processing IV*, pp. 416–30, 1989.

[33] K. Falconer. *Fractal Geometry*. Wiley, 1990.

[34] H. G. E. Hentchel and I. Procaccia. The infinite number of generalized dimensions of fractals and strange attractors. *Physica*, **3D**: 435–44, 1983.

[35] R. F. Brammer. Unified image computing based on fractals and chaos model techniques. *Optical Engineering*, **28**(7): 726–34, 1989.

[36] M. Bertero and Boccacci. *Introduction to Inverse Problems in Imaging*. Institute of Physics Publishing, 1998.

[37] B. Buck and V. A. Macaulay, eds. *Maximum Entropy in Action*. Oxford University Press, 1991.

[38] J. M. Blackledge. *Quantitative Coherent Imaging*. Academic Press, 1989.

[39] M. Turner, J. Blackledge, and P. Andrews. *Fractal Geometry in Digital Imaging*. Academic Press, 1998.

[40] M. Buchanan. One law to rule them all. *New Scientist*, pp. 30–5, November 1997.

[41] R. Hilfer. Scaling theory and the classification of phase transitions. *Modern Physics Letters*, **B6**(13): 773–84, 1992.

[42] P. Bak. *How Nature Works*. Oxford University Press, 1997.

[43] M. F. Shlesinger, M. Zaslavsky and J. Klafter. Strange kinetics. *Nature*, **363**, 31–7, May 1993.

[44] J. Klafter, F. Shlesinger and G. Zumofen. Beyond Brownian motion. *Physics Today*, 33–9, February 1996.

[45] W. Mecklenbräuker and F. Hlawatsch, eds. *The Wigner Distribution: Theory and Applications in Signal Processing*. Elsevier, 1997.

[46] H. Tassignon. Non-stationary deconvolution in wavelet space. Ph.D. thesis, De Montfort University, 1998.

[47] W. H. Press, S. A. Teukolsky, W. T. Vetterling and B. P. Flannery. *Numerical Recipes in C: The Art of Scientific Computing*. Cambridge University Press, 1992.

[48] G. A. Evans, J. M. Blackledge and P. Yardly. *Analytical Solutions to Partial Differential Equations*. Academic Press, 1999.

[49] F. Oberhettinger and L. Badii. *Table of Laplace Transforms*. Springer, 1973.

5 Application to speech processing and synthesis

5.1 Segmentation of speech signals based on fractal dimension

Computer speech recognition is an important subject that has been studied for many years. Until relatively recently, classical mathematics and signal processing techniques have played a major role in the development of speech recognition systems. This includes the use of frequency–time analysis, the Wigner transform, applications of wavelets and a wide range of artificial neural network paradigms. Relatively little attention has been paid to the application of random scaling fractals to speech recognition. The fractal characterization of speech waveforms was first reported by Pickover and Al Khorasani [1], who investigated the self-affinity and fractal dimension for human speech in general. They found a fractal dimension of 1.66 using Hurst analysis (see e.g. [2]). In the present chapter, we investigate the use of fractal-dimension segmentation for feature extraction and recognition of isolated words. We shall start with a few preliminaries that relate to speech recognition techniques in general.

5.1.1 Speech recognition techniques

Speech recognition systems are based on digitizing an appropriate waveform from which useful data is then extracted using appropriate pre-processing techniques. After that, the data is processed to obtain a signature or representation of the speech signal. This signature is ideally a highly compressed form of the original data that represents the speech signal uniquely and unambiguously. The signature is then matched against some that have been created previously (templates) by averaging a set of such signatures for a particular word.

5.1.2 Sampling and data extraction

Any two different samples of the same word may have data that is useful in relation to speech, at different positions along the sampled signal. It is therefore necessary to extract the speech data from the sample so that the processed signals will match correctly. There are several ways to do this. Two of the simplest approaches are based on using:

- the magnitude of the data
- the number of zero crossings in the data

In the first method, if the magnitude of the signal exceeds a certain threshold then the data is used; otherwise it is discarded. The zero-crossing method counts the number of times the signal crosses the zero axis (i.e. the number of times the data changes polarity) for a window of a certain size. If the number of zero crossings is above a certain threshold then the data within the window is used.

The fractal-segmentation technique discussed in this chapter segments a word into windows and processes each window separately. In order for this speech processing technique to work with continuous signals, it is best that the window size used for extracting the data from the original sample is the same as that used for fractal-dimension segmentation.

5.1.3 Template creation

To create a template, several samples for the same word, voiced by different speakers (for voice-independent systems), must be acquired. They are then processed in the same way that the system processes a new input signal. The average of these results forms the word template.

5.1.4 Matching techniques

Three principal methods are used to match the sampled word with existing templates:

- the least squares method (LSM)
- the correlation method (CM)
- dynamic time warping (DTW)

The LSM and CM are standard fitting and matching techniques that are used in many areas of signal and image processing. However, DTW is principally used for speech recognition.

The LSM and CM matching techniques work well when the words compared are of the same length and when corresponding times in separate utterances of a word represent the same phonetic features. In practice, speakers vary their speed of speaking, and often do so non-uniformly, so that different voicings of the same word can have the same total length yet may differ in the middle. For both the least squares and correlation methods, compensation for time-scale variations can be partially made by applying uniform time normalization to ensure that all speech patterns being matched are of the same length. However, this has only limited success and a technique has been developed that is capable of matching one word onto another in a way that applies an optimum non-linear time-scale distortion to achieve the best match at all points. This technique is known as dynamic programming or, when applied to speech signal recognition, dynamic time warping.

Dynamic time warping

If a processed speech sample and template with respective lengths n and N are being compared, a distance can be used to describe the difference $\ell(i, j)$ between the value at i of an incoming speech signal and the value at j of the template.

One way to find the optimum path is to evaluate the sum of $\ell(i, j)$ for all possible paths. This method will always give the correct answer but is prohibitively computationally expensive. Dynamic programming allows for identification of the optimum path without calculating the lengths of all possible paths.

Assuming that the path goes forward with a non-negative slope, as would be expected, and that the present point $L(i, j)$ is on the optimum path then the preceding point must be one of $(i-1, j)$, $(i-1, j-1)$ and $(i, j-1)$. If $L(i, j)$ is the cumulative distance along the optimum path from the beginning of the word to point (i, j) then

$$L(i, j) = \sum_{(x,y)=(i,j)} \ell(x, y)$$

where the sum is taken along the optimum path. As there are only three points that could be positions previous to (i, j) it is clear that

$$L(i, j) = \min[L(i-1, j), L(i-1, j-1), L(i, j-1)] + \ell(i, j)$$

Now, $L(1, 1)$ must be equal to $\ell(1, 1)$ as this is at the start of all possible paths. Using $L(1, 1)$ as the start, the values in the vertical column $L(1, j)$ can be calculated using a reduced form of the above equation. When this column of values is known, the full equation can be applied for columns $2, \ldots, n$. A measure of the difference between the two signals is given by the value calculated for $L(n, N)$.

5.2 Isolated word recognition

In order to test the validity and accuracy of the different methods for computing the fractal dimension, a fractal signal of size 256 with a known fractal dimension D was created by filtering white Gaussian noise with the filter $1/k^q$, where $q = (5 - 2D)/2$. The fractal dimensions of the synthesized fractal signal were then evaluated using the walking-divider, box-counting and power spectrum methods. The results are given in Table 5.1. From this table, it is clear that the power spectrum method (PSM) provides the most consistently accurate results throughout the range [1, 2]; this is to be expected as the data used to undertake this test was generated using this method. The box-counting method provides good results for fractal dimensions with a value below 1.5. After this, the fractal dimension is below the original value used to synthesize the data; for a value of 2.0, the box-counting method returns a value of 1.6. However, this does not mean that the box-counting method will give results as low as this for all the fractal signals used, as the test procedure considered here utilizes a signal generator

Table 5.1 *Evaluation and comparison of fractal dimensions*

Original value of D	WDM	BCM	PSM
1.0	1.220	1.097	1.006
1.1	1.259	1.147	1.138
1.2	1.294	1.165	1.219
1.3	1.351	1.261	1.273
1.4	1.404	1.307	1.382
1.5	1.451	1.380	1.495
1.6	1.501	1.418	1.599
1.7	1.542	1.482	1.705
1.8	1.592	1.537	1.832
1.9	1.617	1.554	1.942
2.0	1.604	1.561	1.997

based on filtering white Gaussian noise. The walking-divider method provides a good approximation of the fractal dimension for values below 1.5, returning results that are slightly higher than those produced by the box-counting method. For an original value of 2.0 the walking-divider method returns a value of 1.604.

Of the three methods tested the PSM is the fastest; this is as expected since it is based on a non-iterative approach using a least squares estimate that relies on the use of an FFT, as considered in Section 5.2. The accuracy, efficiency and versatility of the PSM leads naturally to its use in many areas of signal and image processing. The test discussed above gives confidence that the PSM is the most appropriate technique for applications to speech processing and synthesis.

5.2.1 Pre-processing data

Direct application of the PSM for computing the fractal dimension of arbitrary speech signals leads to a wide range of values, many of which lie outside the range $[1, 2]$. This is not surprising, since many speech waveforms will not conform to patterns that are statistically self-affine with a single spectral signature of the type $1/k^q$. It is expected that, as with many other signals, some aspects of speech are fractal in nature while other aspects are not. In other words, like any other model for signal analysis, a fractal model cannot be assumed to be applicable to all aspects of speech. As with any other signal, one should expect speech to be composed of both fractal and non-fractal components, particularly since highly non-stationary waveforms are observed in speech. Some pre-processing is therefore required in order to force a speech signal or component waveform to conform to a spectral signature of a fractal type, in particular, some appropriate low-pass filter. Of all the possible low-pass filters, a filter of the type $1/k$ is the most appropriate as it conforms to the case $q = 1$. All data is then constrained to appear as $1/k^q$ noise.

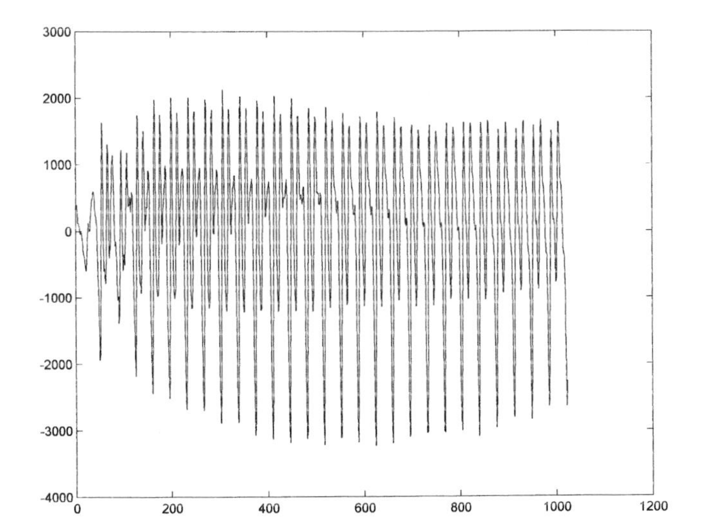

Figure 5.1 Waveform for the fricative 'o'; $D = 1.20$. The magnitude of the signal is plotted against time elapsed.

In terms of the stochastic fractal differential equation

$$\frac{d^q}{dx^q} f(x) = n(x)$$

where n is white noise, this pre-filtering operation is equivalent to modelling a fractal speech signal f in terms of the modified equation

$$\frac{d^q}{dx^q} f(x) = \int^x n(x)\, dx$$

since the integral on the right-hand side is equivalent to application of the integrator filter $1/ik$. This approach leads to a segmentation method that, in effect, is based on measuring the behaviour of the waveform in terms of its deviation from the case $q = 1$.

Evaluation

The following results are based on a single male speaker and the fricatives 'o', 'z' and 'sh'. The results illustrate the expected increase in frequency content and zero crossings (as illustrated by direct inspection of the waveforms) and the increasing value of the fractal dimension. Figure 5.1 shows the waveform for 'o', which, through application of the PSM with $1/k$ pre-filtering, returns a fractal dimension 1.20. Figure 5.2 shows the waveform for the fricative 'z', which is characterized by a fractal dimension 1.38. Finally, Figure 5.3 shows the waveform for the fricative 'sh', which yields a fractal dimension 1.58.

An example from which the fractal dimension for a single isolated word can be computed is given in Figure 5.4, which shows the waveform for the word 'test', spoken by a female speaker. Application of the power spectrum method with $1/k$ pre-filtering

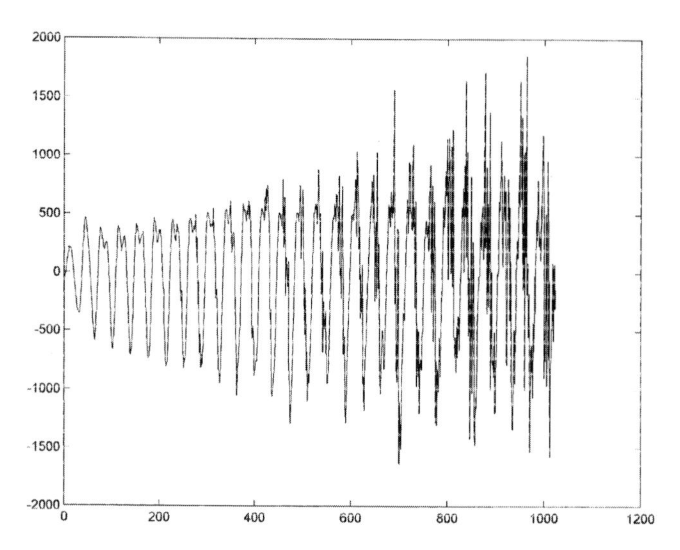

Figure 5.2 Waveform for the fricative 'z'; $D = 1.38$.

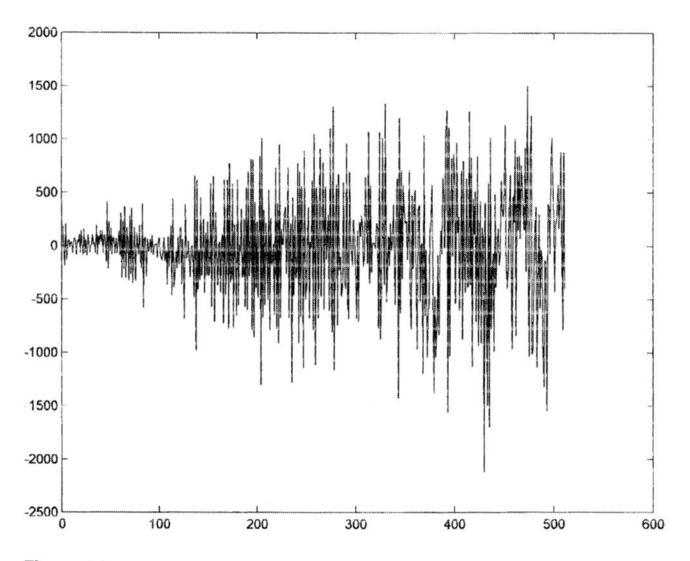

Figure 5.3 Waveform for the fricative 'sh'; $D = 1.58$.

for this signal returns a value 1.34 for the fractal dimension. Figures 5.5–5.8 show the waveforms that characterize the fricative components of this signal for the word 'test', namely 't', 'e', 's' and 't', which return values 1.06, 1.36, 1.41 and 1.19 respectively for the fractal dimension (waveforms for the word 'test' were introduced in Chapter 1). Note that the fractal dimension for 't' at the beginning and at the end of the word is different, owing to changes in the pronunciation of this fricative when used to form a complete word. Also note that, as expected, the fractal dimension is greatest for the high-frequency fricative 's'. This result is a simple illustration of the way in which

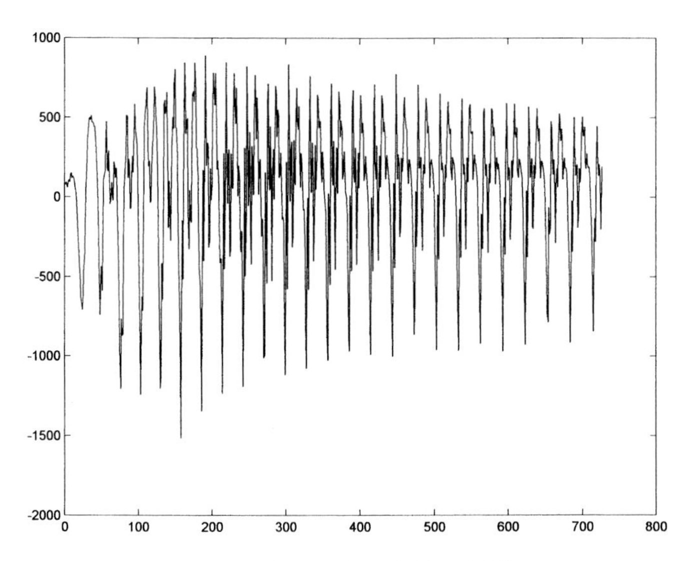

Figure 5.4 Waveform for the word 'test'; $D = 1.34$.

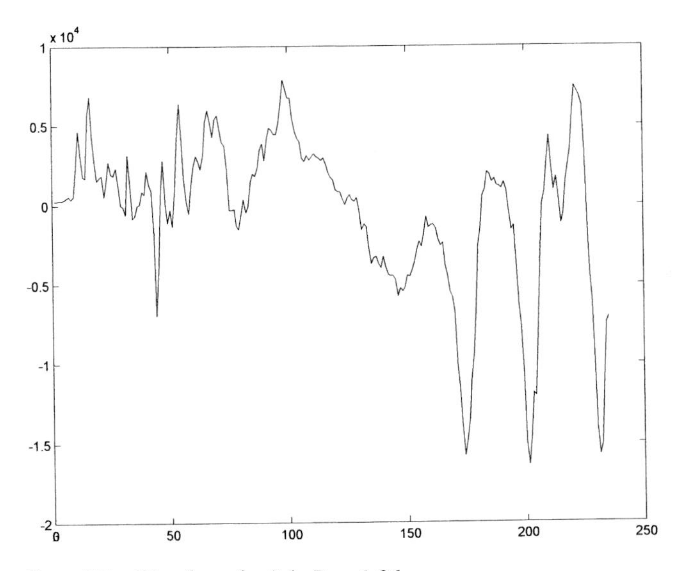

Figure 5.5 Waveform for 't '; $D = 1.06$.

the waveforms associated with isolated words can be segmented into a set of fractal dimensions, each representative of the individual components of the word.

5.2.2 Generalization of the RSF model

Although statistical self-similarity and self-affinity are properties of many signals found in nature, the power spectrum density function (PSDF) associated with a fractal signal (i.e. the $k^{-\beta}$ model) is not appropriate to all noise types and/or the whole spectrum.

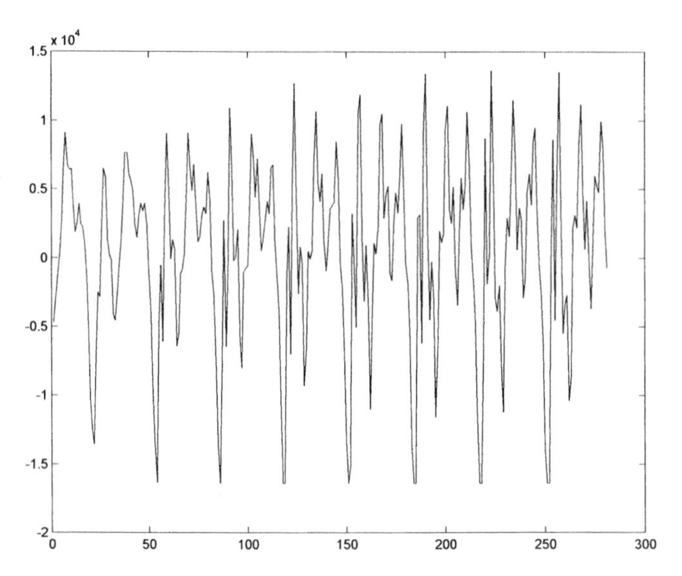

Figure 5.6 Waveform for 'e'; $D = 1.36$.

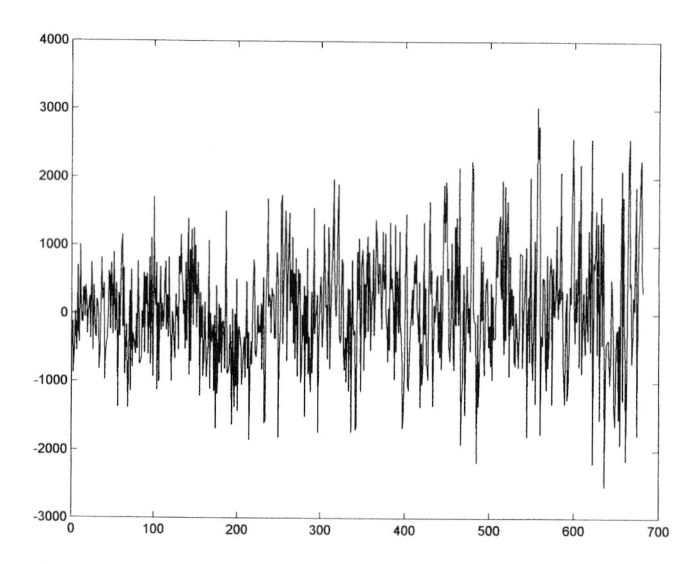

Figure 5.7 Waveform for 's'; $D = 1.41$.

Many signals do have a high-frequency decay, for which the fractal model is appropriate, but the complete power spectrum may have characteristics for which a simple $1/k^q$ power law is not appropriate. This has led to the development of spectral partitioning algorithms, which attempt to extract the part of the spectrum to which the $1/k^q$ power law applies most appropriately. Alternatively, $1/k$ pre-filtering can be applied to force the PSDF to conform to a $1/k^q$ scaling law, as described in the previous subsection with regard to processing speech signals.

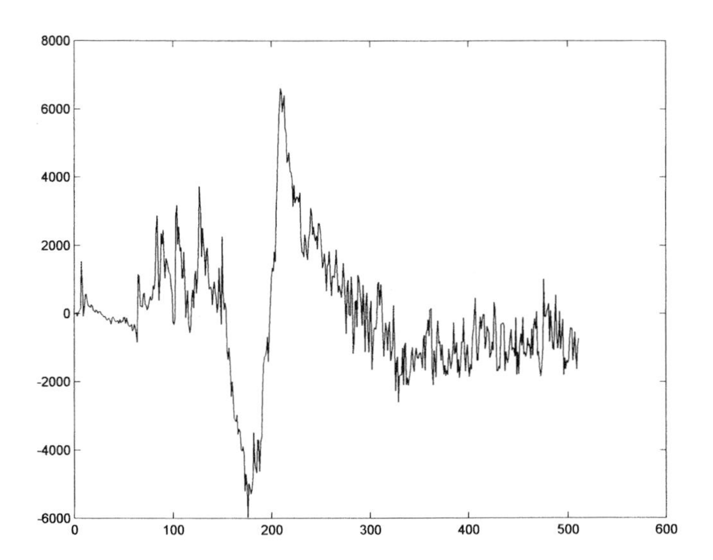

Figure 5.8 Waveform for 't'; $D = 1.19$.

Is there a more general model that can be used to characterize a wider variety of PSDFs, of which the $k^{-\beta}$ model is a special case? Following Tatom [3], we consider the PSDF

$$P(k) = \frac{c|k|^{2g}}{(k_0^2 + k^2)^q} \tag{5.1}$$

where g and q are positive (floating-point) numbers, c is a scaling factor and k_0 is a characteristic frequency. Clearly, this model reduces to the PDSF for a fractal signal when $g = 0$, $k_0 = 0$ and $1 < q < 2$, but it also includes other stochastic models such as the Ornstein–Uhlenbeck process [4], in which the corresponding PSDF is

$$P(k) = \frac{ck}{k_0^2 + k^2}$$

and the Bermann process [5], where

$$P(k) = \frac{c|k|^g}{k_0^2 + k^2}$$

Use of the Ornstein–Uhlenbeck process stems from the difficulties of non-differentiability associated with a Wiener process. The process is Gaussian and Markov but does not have stationary increments. It is the only example known of a Gaussian process that is stationary and Markov.

For scalar fields defined in an n-dimensional space, the fractal dimension associated with the Bermann process is given by

$$D = \min\left[\frac{2n}{1-g}, \; n + \frac{1+g}{2}\right]$$

The properties of self-similarity and self-affinity do carry over to these more general processes but in a more restrictive way.

5.2.3 Basic properties

For $k > 0$, the function $P(k)$ given in Eqn 5.1 has a maximum when

$$\frac{\mathrm{d}}{\mathrm{d}k}\ln P(k) = \frac{2g}{k} - \frac{2kq}{k_0^2 + k^2} = 0$$

i.e. when

$$\frac{\mathrm{d}}{\mathrm{d}k}P(k) = \left(\frac{2g}{k} - \frac{2kq}{k_0^2 + k^2}\right)P(k) = 0$$

Since $P(k) \neq 0 \; \forall \; k$, this implies that the maximum value of $P(k)$ occurs at a value $k = k_{\text{max}}$ given by

$$k_{\text{max}} = k_0\sqrt{\frac{g}{q-g}}, \qquad q > g$$

The value of $P(k)$ at this point is therefore

$$P_{\text{max}} \equiv P(k_{\text{max}}) = \frac{ck_{\text{max}}^{2g}}{(k_0^2 + k_{\text{max}}^2)^q} = ck_0^{2(g-q)}\frac{g^g}{q^q}(q-g)^{q-g}$$

Beyond this point, the PSDF decays and its asymptotic form is dominated by a $k^{-\beta}$ power law, which is consistent with RSF signals and many noise types at the high-frequency end of their power spectrum. At low frequencies, the spectrum is characterized by the term $(ik)^g$.

5.2.4 Analysis

Given Eqn 5.1, the complex spectrum of the noise $F(k)$ can be written as

$$F(k) = H(k)N(k)$$

where as before $N(k)$ is the complex spectrum of white (δ-uncorrelated) noise and $H(k)$ is the 'transfer function' (ignoring scaling):

$$H(k) = \frac{(ik)^g}{(k_0 + ik)^q}$$

The noise function $f(x)$ is then given by

$$f(x) = \frac{1}{2\pi}\int_{-\infty}^{\infty} H(k)N(k)\exp(ikx)\,\mathrm{d}k$$

It is interesting to analyse this result further with the aim of establishing the transform of $n(x)$ and so obtain $f(x)$. If we consider the definition of a fractional derivative in terms of the inverse Fourier transform of $(ik)^g$ then, using the convolution theorem, we can write

$$f(x) = \int h(x-y)\frac{\mathrm{d}^g}{\mathrm{d}y^g}n(y)\,\mathrm{d}y$$

where

$$h(x) = \frac{1}{2\pi}\int_{-\infty}^{\infty}\frac{1}{(k_0+ik)^q}\exp(ikx)\,\mathrm{d}k$$

Substitution of p for ik allows us to write this result in terms of the inverse Laplace transform, i.e.

$$h(x) = \hat{L}^{-1}\left[\frac{1}{(k_0+p)^q}\right]$$

Since

$$\hat{L}[x^q\exp(-k_0x)] = \frac{\Gamma(q+1)}{(k_0+p)^{q+1}}, \qquad q>-1, \quad \mathrm{Re}(p+k_0)>0$$

it follows that

$$h(x) = \frac{1}{\Gamma(q)}\frac{\exp(-ik_0x)}{x^{1-q}}$$

Hence $f(x)$ can be written in terms of the fractional integral transform:

$$f(x) = \frac{1}{\Gamma(q)}\int_{-\infty}^{x}\frac{\exp[-k_0(x-y)]}{(x-y)^{1-q}}\frac{\mathrm{d}^g}{\mathrm{d}y^g}n(y)\,\mathrm{d}y$$

The scaling characteristics of this transform can be investigated by considering the function

$$\begin{aligned} f'(x,k_0) &= \frac{1}{\Gamma(q)}\int_{-\infty}^{x}\frac{\exp[-k_0(x-y)]}{(x-y)^{1-q}}\frac{\mathrm{d}^g}{\mathrm{d}y^g}n(\lambda y)\,\mathrm{d}y \\ &= \frac{\lambda^g}{\lambda^q}\frac{1}{\Gamma(q)}\int_{-\infty}^{\lambda x}\frac{\exp[-k_0/\lambda(\lambda x-z)]}{(\lambda x-z)^{1-q}}\frac{\mathrm{d}^g}{\mathrm{d}z^g}n(z)\,\mathrm{d}z \\ &= \frac{\lambda^g}{\lambda^q}f\left(\lambda x, k_0/\lambda\right) \end{aligned}$$

on substitution of z for λy. Hence, the scaling relationship for this model is

$$\Pr[f'(x,k_0)] = \frac{\lambda^g}{\lambda^q}\Pr[f(\lambda x, k_0/\lambda)]$$

Here, as we scale x by λ the characteristic frequency k_0 is scaled by $1/\lambda$ – a result that is in some sense consistent with the scaling property of the Fourier transform (i.e. $f(\lambda x) \Longleftrightarrow 1/\lambda F(k/\lambda)$). The interpretation of this result is that as we zoom into the signal $f(x)$, the distribution of amplitudes, i.e. the probability density function, remains the same (subject to a scaling factor λ^{g-q}) and the characteristic frequency of the signal increases by a factor $1/\lambda$.

Clearly, all the results discussed above reduce to the 'normal' theory of RSF signals when $g = 0$ and $k_0 = 0$, but this model provides a much greater degree of flexibility in terms of characterizing the PSDFs of many noise types, nearly all of which have some degree of statistical self-affinity, and PSDFs with power laws of irrational form.

In using this model to characterize texture, we consider the case where a suitable combination (some cluttering algorithm) of the parameters g, q, k_0 and c is taken to be a measure of texture, in particular, the parameters g and q – their product for example. In such a case we are required to obtain estimates for these parameters associated with the data $f_i, i = 1, 2, \ldots, N$. The general four-parameter problem is not easily solved, primarily because of difficulties in linearizing $P(k)$ with respect to k_0. However, suppose that a good estimate for k_0 can be obtained; then we can compute estimates for g, q and c using a standard least squares method by constructing a logarithmic least squares estimate in the usual way, i.e. we consider (with $C = \ln c$)

$$e(g, q, C) = \sum_i (\ln P_i - \ln \hat{P}_i)^2$$

where P_i is the discrete power spectrum of f_i and

$$\hat{P}_i = \frac{c|k_i|^{2g}}{(k_0^2 + k_i^2)^q}$$

is its expected form. In this case,

$$e(g, q, C) = \sum_i \left[\ln P_i - 2g \ln |k_i| - C + q \ln(k_0^2 + k_i^2)\right]^2$$

which is a minimum when

$$\frac{\partial e}{\partial g} = 0, \qquad \frac{\partial e}{\partial q} = 0, \qquad \frac{\partial e}{\partial C} = 0$$

Differentiating, it is easy to show that the parameter set (g, q, C) is given by the solution to the following linear system of equations:

$$\begin{pmatrix} a_{11} & a_{21} & a_{31} \\ a_{12} & a_{22} & a_{32} \\ a_{13} & a_{23} & a_{33} \end{pmatrix} \begin{pmatrix} g \\ q \\ C \end{pmatrix} = \begin{pmatrix} b_1 \\ b_2 \\ b_3 \end{pmatrix}$$

where

$$a_{11} = -\sum_i (\ln |k_i|)^2$$

$$a_{21} = \sum_i \ln(k_0^2 + k_i^2) \ln |k_i|$$

$$a_{31} = -\sum_i \ln |k_i|$$

$$a_{12} = -\sum_i (\ln |k_i|)\left[\ln(k_0^2 + k_i^2)\right]$$

$$a_{22} = \sum_i \left[\ln(k_0^2 + k_i^2)\right]^2$$

$$a_{32} = -\sum_i \ln(k_0^2 + k_i^2)$$

$$a_{13} = -\sum_i \ln |k_i|$$

$$a_{23} = \sum_i \ln(k_0^2 + k_i^2)$$

$$a_{33} = -N$$

$$b_1 = -\sum_i (\ln P_i)(\ln |k_i|)$$

$$b_2 = (\ln P_i)\left[\ln(k_0^2 + k_i^2)\right]$$

$$b_3 = -\sum_i \ln |P_i|$$

An initial estimate for k_0 can be obtained from the result

$$k_0 = k_{\max}\sqrt{\frac{q-g}{g}}, \qquad q > g \tag{5.2}$$

where $k_{\max}$ is the frequency corresponding to the maximum value of the power spectrum. The value of $k_{\max}$ can be estimated by applying a smoothing process to the power spectrum (a moving-average filter, for example) and then computing the mode of the resulting distribution. Having obtained an estimate for $k_{\max}$, we consider the case $k_0 = k_{\max}$, which will give a first approximation to the parameter set (g, q, C). An iteration procedure can then be adopted in which the initial estimates of g and q are used to compute a new value for k_0 via Eqn 5.2 and the linear system of equations given above solved to obtain a second estimate of the parameter set (g, q, C), and so on.

5.2.5 Conclusions and discussion

The use of texture segmentation in digital signal and image processing remains a rather speculative subject. It is difficult at present to state which measure or combination of measures of texture is best suited to a particular class of data for a given application. The application of the fractal dimension as a measure of texture has a number of important implications but fundamentally assumes that the data is statistically self-affine.

On the one hand, a common criticism concerning the use of fractals in signal processing is that signal analysis should proceed using the measured statistics and not on the assumption that the signal obeys some model (fractal or otherwise). On the other hand, if a model is assumed which provides results that are consistent with its theoretical basis and which, in addition, provides a useful segmentation method for automated feature extraction and pattern recognition then a model-based approach may be desirable.

In the work reported here, application has been made of a well-tested algorithm for computing the fractal dimension of a fractal signal for speech processing. In numerical work carried out so far, all the speech data used has consistently provided fractal dimensions in a valid physical range providing that the data is appropriately pre-filtered. It has been found that a pre-filter of the form $1/k$ is particularly well suited for the fractional-dimension segmentation (FDS) of speech signals.

The application of RSF models is not entirely appropriate to all forms of signal analysis and considerable effort has been given to classifying signals with genuine fractal characteristics. This is usually done by considering the spectral characteristics of a signal and observing whether they are consistent with a $k^{-\beta}$ model. In this book, a more general model has been investigated where the PSDF of a stochastic signal is assumed to be of the form (ignoring scaling) $|k|^{2g}(k_0^2 + k^2)^{-q}$. This model encompasses the fractal model and other stochastic processes such is the Ornstein–Uhlenbeck and Bermann processes. We propose that the parameters g and q should be used for texture segmentation on a wider class of signals and images than is appropriate for FDS alone [6].

REFERENCES

[1] C. A. Pickover and A. Al Khoransani. Fractal characterization of speech. *Journal of Computer Graphics*, **10**(1): 51–61, 1986.

[2] M. Nakagawa. A critical exponent method to evaluate fractal dimensions of self-affine data. *J. Physical Society of Japan*, **62**(12), 1993.

[3] F. B. Tatom. *The Application of Fractional Calculus to the Simulation of Stochastic Processes*, vol. AIAA-89/0792. Engineering Analysis Inc., Huntsville, Alabama, 1989.

[4] D. Cox and H. Miller. *The Theory of Stochastic Processes*, pp. 205–8. Chapman and Hall, 1972.

[5] R. Adler. *The Geometry of Random Fields*. Springer, 1990.

[6] M. Turner, J. Blackledge and P. Andrews. *Fractal Geometry in Digital Imaging*. Academic Press, 1998.

6 Speech processing with fractals

6.1 Introduction

A signal is defined as any physical quantity that varies with time, space or any other independent variable or variables. Mathematically, we describe a signal as a function of one or more independent variables. For example, the functions

$$S_1(t) = 5t \tag{6.1}$$

$$S_2(t) = 20t^2 \tag{6.2}$$

describe two signals, one that varies linearly with the independent variable t (time) and a second that varies quadratically with t. As another example, consider the function

$$S(x, y) = 3x + 2xy + 10y^2 \tag{6.3}$$

This signal function has two independent variables x and y, which might represent the two spatial coordinates in a plane.

The signals described by Eqns 6.1 and 6.2 belong to a class of signals that are precisely defined by specifying the functional dependence on the independent variable. However, there are cases where such a functional relationship is unknown or too highly complicated to be of any practical use. For example, the speech signal depicted in Figure 6.1 cannot be described functionally by expressions such as in Eqn 1. In general, a segment of speech can be represented to a high accuracy as a sum of several sinusoids of different amplitudes and frequencies,

$$\sum_{i=1}^{N} A_i(t) \sin\left[2\pi F_i(t)t + \theta_i(t)\right] \tag{6.4}$$

where $A_i(t)$, $F_i(t)$ and $\theta_i(t)$ are the (time-varying) amplitude, frequency and phase respectively of sinusoidi i. In fact, one way to interpret the information content or message conveyed by any short-time segment of the speech signal is to measure the amplitudes, frequencies and phases contained in the short-time segment.

Another example of a natural signal is an electrocardiogram. Such a signal provides a doctor with information about the operation of the patient's heart. Similarly,

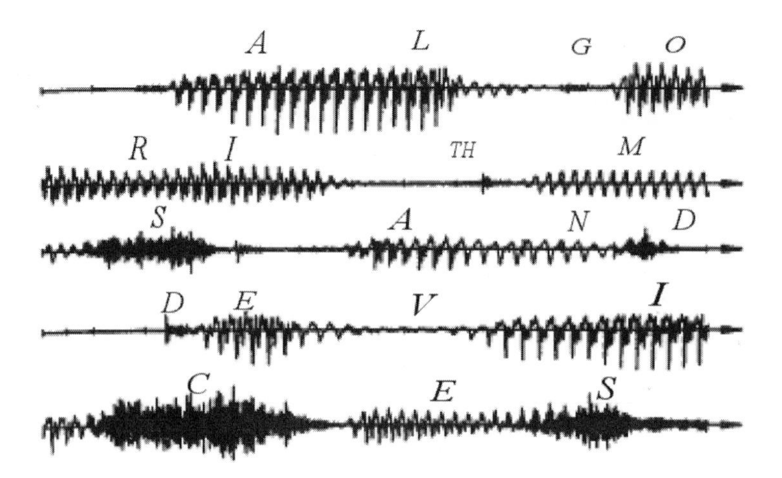

Figure 6.1 Examples of speech signals.

an electroencephalogram signal provides information about the activity of the brain. Speech, electrocardiogram and electroencephalogram signals are examples of information-bearing signals that evolve as functions of a single independent variable, namely, time. An example of a signal that is a function of two independent variables is an image signal. The independent variables in this case are the spatial coordinates. These are but a few examples of the countless numbers of natural signals that are encountered in practice.

Associated with the natural (i.e. original) signal is the means by which such a signal is generated. For example, speech signals are generated by forcing air through the vocal chords (vocal folds). Images are obtained by exposing a photographic film to a scene or an object. Thus signal generation is usually associated with a *system* that responds to a stimulus or force. In a speech signal, the system consists of the vocal chords and the vocal tract, also called the vocal cavity. The stimulus in combination with the system is called a signal source. Therefore, we have speech sources, image sources and various other types of signal source.

A system may also be defined as a *physical device* that performs an operation on a signal. For example, a filter that is used to reduce the noise and interference corrupting a desired information-bearing signal is called a system. In this case the filter performs an operation or operations on the signal; this has the effect of reducing (filtering) the noise and interference from the desired information-bearing signal. When a signal passes through a system it can be said that the signal is being processed. In this case the processing of the signal involves filtering the noise and interference from the desired signal. In general, the system is characterized by the type of operation that it performs on the signal. For example, if the operation is linear, the system is called linear. If the operation on the signal is non-linear the system is said to be non-linear, and so forth. Such operations are usually referred to as *signal processing* [1].

Speech processing by computer provides one vehicle for 'natural' communication between people and machines. Although fluent communication in spoken natural English for unrestricted applications is not possible at the present, a great deal is known about how to build computer systems which recognize and in a formal sense 'understand' a restricted range of spoken commands and which can synthesize intelligible speech output. Moreover, the foundations have been laid, and research is in progress, to develop technology that will permit natural English communication in continuous speech to be used for an increasingly complex range of applications. With the increasing role of computers in modern life, there is little doubt that, as well as talking to them we will want them to talk to us. It is the science and technology necessary for the construction of such machines [2] that is lacking.

This chapter presents what is known about the nature of speech and speech processing technology today and portrays some of the directions that future research will take. It is not intended to be a first introduction to speech processing, although it contains enough tutorial material to serve as such for a dedicated reader. Rather, it is intended as an advanced text that will prepare a student for further study, leading to a role in speech research or to the use of these techniques in computer applications. It should also be useful to someone who needs to understand the capabilities and limitations of speech processing technology at a technical level.

Research in speech processing by computer has traditionally been focused on a number of somewhat separable but overlapping problems. One of these is isolated word recognition, where the signal to be recognized consists of a single word or phrase, delimited by silence, that is to be regarded as a unit without characterization of its internal structure. For this kind of problem, certain traditional pattern-recognition techniques can be applied directly. Another area is speech compression and re-synthesis, where the objective is to transform the speech signal into a representation that can be stored or transmitted with a smaller number of bits of information and later restored to an intelligible speech signal. Here, a number of powerful signal processing and information theoretic techniques are available.

A more difficult problem is the synthesis of speech by rule from an initial textual representation. Here, the system must deal with the phonological rules that govern the pronunciation of words and the co-articulation effects that occur between words, as well as the intonation, rhythm and energy contour (collectively referred to as *prosodics*) that must be imposed on the utterance in order for the result to be intelligible.

The most difficult of the speech processing tasks is the understanding of continuous spoken utterances from unknown speakers using fluent English syntax. Here, the system needs to deal with the inverses of most of the difficult issues of speech synthesis, and it also has the problem of searching for viable interpretations of locally ambiguous phenomena. In addition one may be dealing with not just one voice and one intonation strategy but an unknown voice and any one of many different possible intonation strategies that the speaker might have employed. In addition to acoustic, phonetic and prosodic information one needs to draw on knowledge of the grammar and phonology of English, dialect and individual speaker differences and pragmatic knowledge of what kinds of utterance make sense in different circumstances. It is also important to know the ways in which the syntax meaning and purpose of an utterance affect its prosodics.

Due to the role of speech in intelligent communication, speech recognition can serve as a focus for research in a number of disciplines, including acoustics, signal processing, linguistics, cognitive psychology and artificial intelligence. In its most comprehensive form, speech recognition draws on results from all of these areas.

Much of speech recognition research in the past has been conducted in isolation from related work in phonology, syntax and general theories of intelligence; however, research in the 1990s began to reveal a more comprehensive picture of how human speech participates as a component of larger patterns of intellectual activity. One of the major catalysts of this larger picture has been the confluence of traditional speech research with advances in digital signal processing techniques and the results of research in computational linguistics and artificial intelligence. This has brought to bear on the problem of speech recognition a new level of detailed understanding of both the low-level physical processes of speech production and the higher-level linguistic and cognitive activities that play critical roles in formulating and understanding speech utterances.

Fractal signature

The application of fractal geometry to speech signals and speech recognition systems is now receiving serious attention, as mentioned in earlier chapters. A very important characteristic of fractals, useful for their description and classification, is their fractal dimension D [3]. The fractal dimension provides an objective means of quantifying the fractal property of an object and comparing objects observed in the natural world. Fractals thus provide a simple description of many natural forms. Intuitively D measures the degree of irregularity over multiple scales. The application of fractal theory depends on the accurate measurement of D – this is essential in fractal-based speech recognition. Fractal speech recognition can be generally defined as the process of transforming continuous acoustic speech signals into a discrete representation. It involves the identification of specific words, by comparison with stored templates. Fractal signatures for various utterances were discussed in subsection 5.2.1.

6.2 Sampling strategies for speech

Signal analysis has a central role in speech processing. It is used in phonetics, for example in the production of spectrograms. It is used in communication channels, where its parameterization allows the data rate to be reduced compared with that required for the transmission of directly digitized speech. It is used in speech synthesis where

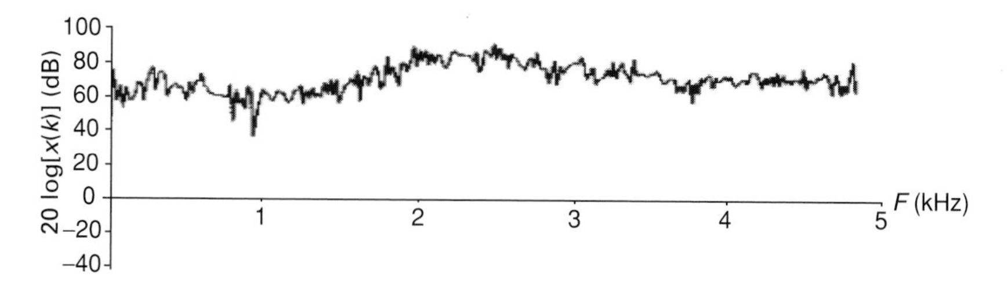

Figure 6.2 Spectrum of 'a' sound.

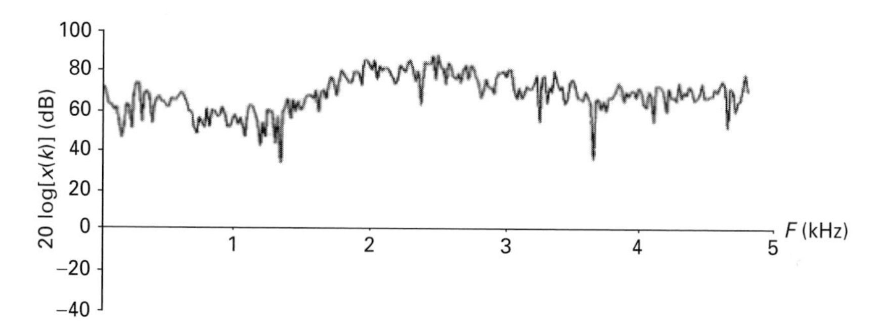

Figure 6.3 Spectrum of 's' sound.

pre-analysed or parameterized segments of speech are reconstructed back into speech. It is used in speech recognition, as a first part or front end of the process with the task of providing parameters that are used as data for the recognition process.

An important parameter in speech analysis is the frequency range of speech. The spectra of the voiced sound 'a' and the unvoiced sound 's' are shown in Figures 6.2 and 6.3 respectively. The frequency bandwidth of a speech signal is about 16 kHz [4]. However, most speech energy is under 7 kHz. Speech bandwidth is generally reduced in recording. A speech signal is called *orthophonic* if all the spectral components over 16 kHz are discarded. A telephonic lower-quality signal is obtained whenever a signal does not have energy outside the band 300–3400 Hz. Therefore, digital speech processing is usually performed by a frequency sampling ranging between 8000 samples per second and 3200 samples per second. These values correspond to bandwidths of 4 kHz and 16 kHz respectively.

Another important and related parameter in speech processing is the *sampling rate*. In any digital processing the original waveform, such as a speech signal, is a continuous analogue quantity, say $x_a(t)$, but since a computer carries out its operations on discrete quantities, the analogue quantity has to be sampled to give discrete quantities by an analogue-to-digital converter (ADC). Thus the analogue signal $x_a(t)$ is sampled into a discrete sequence of values $x(0), x(T), x(2T), \ldots$ at a sampling rate of T^{-1} per second. We commonly leave out the T and describe sequences as $x(0), x(1), \ldots$ or as $x(n)$, with the general value $x_a(n)$.

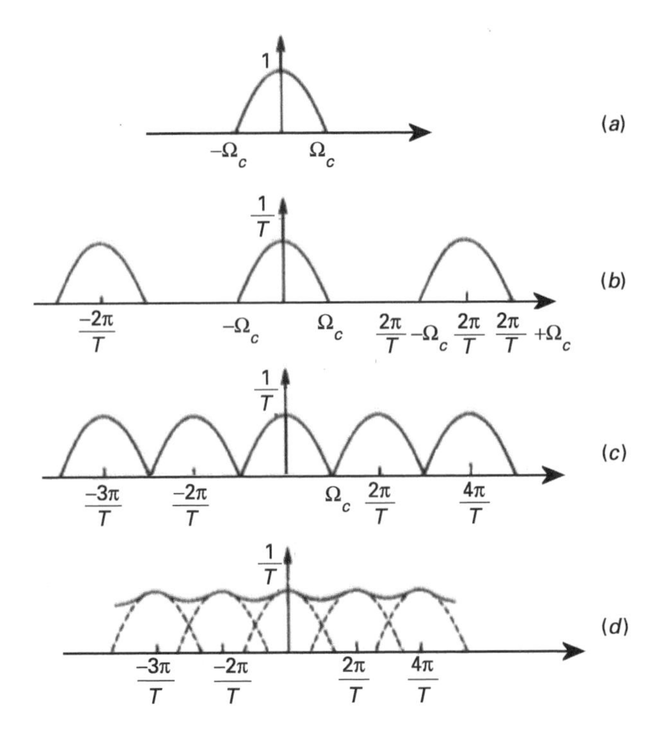

Figure 6.4 Spectra $x(e^{i\Omega t})$ of sampled signals. $\Omega_c = 2\pi f_c$.

There is an important relationship between sampling rate and signal bandwidth. It was shown in [4] that if a band-limited signal, whose frequency components are zero above a frequency f_c, is sampled at a rate $F_s = T^{-1}$ then the spectrum of the sampled signal, $x(e^{i\Omega t})$, is periodic with frequency F_s and of the form shown in Figure 6.4.

Figure 6.4 shows (*a*) the spectrum of the analogue signal, (*b*) the spectrum of the sampled signal for a relatively high sampling rate $F_s \gg f_c$, (*c*) the spectrum for a sampling rate $F_s = f_c$ and (*d*) the spectrum for a relatively low sampling rate $F_s < 2f_c$. It is seen that for high-rate sampling the periodic components of the spectrum are distinct, whereas at $F_s = 2f_c$ they touch and at low rates $F_s < 2f_c$ they overlap. The critical sampling rate f_c is termed the *Nyquist frequency*. An important consequence is that, for sampling rates greater than or equal to the *Nyquist rate* $2f_c$, the original analogue signal can be recovered from its sampled version, whereas sampling rates below the Nyquist rate cause 'aliasing' of the signal.

Thus when speech at an effective bandwidth of 5 kHz is to be sampled it must first be band-limited to 5 kHz by filtering the analogue signal to remove the higher-frequency components and then sampled at a frequency of at least 10 kHz.

Another important parameter in speech sampling is the quantization of the amplitude caused by using an ADC with its limited resolution. Thus for example an 8 bit ADC can convert an input signal into one of 256 levels. This introduces noise into the measurement. Quantization noise can be reduced by increasing the resolution of the

ADC. For speech processing it is common to use an ADC of at least 10 bits, giving a resolution of one part in 1024.

6.3 Numerical algorithms and examples

Let us start by describing an analogue signal. The word 'analogue' implies a resemblance or similarity between one thing and another. An acoustic signal, the movement of molecules in a medium such as air, can be transformed into a voltage signal that travels down an electrical wire. The voltage in the wire is an analogue or representation of the acoustic signal. A typical setup may involve a musical instrument that creates a tone. This tone energy creates a disturbance in the air particles by pushing air molecules, which are compressed; a rarefaction is caused when the air expands from the compression region. This movement is happening at a fast rate that equals that of the initial source of the sound wave. This tone is then received by a microphone, which has a sensitive diaphragm that responds to the movement of air. It is known as a *transducer*, because it converts the energy from an acoustic signal to an electrical signal that represents the same waveform. This voltage signal is then carried down a wire to an amplifier, where the signal is amplified and sent down another wire to a loudspeaker, which transforms the signal back to acoustic energy, which is finally received by the auditory system. Filters, ring modulators, simple frequency modulation, and sample and hold systems are modifiers that may be used to alter an analogue signal.

In the digital world, numbers are used to represent a sampled waveform. Thus, an audio signal is represented in digital memory with a binary code that stores the massive amount of numbers that are needed to represent a signal. As discussed in the previous section, an ADC is a computer chip that converts an analogue signal (Figure 6.5) into digital information. This sampling process has changed the world of sound representation in a dramatic fashion.

Once a signal is represented as binary numbers, it may be manipulated with processes involving combining sounds, splicing, truncation, looping, reversing a sound or other digital signal processing. The signal is then converted back to an analogue signal through a digital-to-analogue converter (DAC).

Figure 6.5 Acoustic waveform.

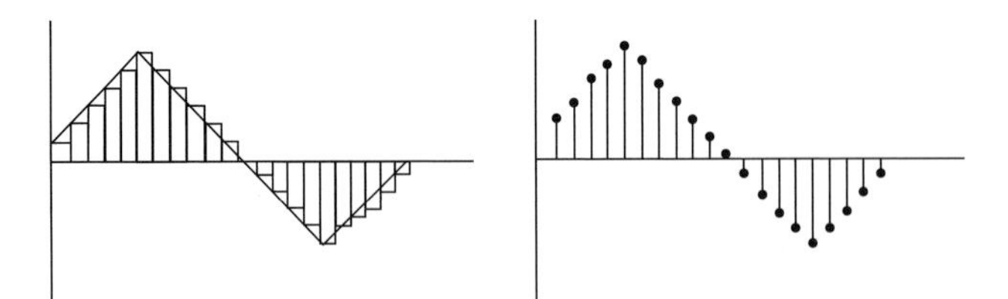

Figure 6.6 On the right, digital sampling of the waveform shown on the left.

Analogue-to-digital conversion

In order to sample a sound event we must understand the process used to convert the acoustic waveform to a digital sample, as shown in Figure 6.6. for a triangular waveform. The user has to determine the sampling period or rate and must also decide on the bit resolution. Let us use a sampling period of one second. The sampling rate determines the number of samples that will be taken during that one second. The bit resolution determines where a signal sample will be represented. If the resolution were at 16 bits then there would be 65 536 locations to represent the waveform for each given sample. That would be a range from −32 768 through 0 to 32 767. If the resolution were at 8 bits then there would be 256 possible locations. The term *quantizing* refers to the process whereby the actual sample is shifted in value to one of the bit locations that best represents the signal at that discrete moment in time. If we change the sampling rate, the period of time or space between each sample is changed. In Figure 6.7, four examples of different quantizations (bit resolutions) and sampling rates, are given. In Figure 6.7(*a*), (*b*), the sampling rate is the same, but the quantization (bit resolution) is better in (*b*). In (*c*) and (*d*) the sampling rate has been doubled, and furthermore in (*d*) the bit resolution has been increased again. What a dramatic difference the sampling rate and bit resolution can make when later recreating an acoustic waveform!

The *Nyquist theorem* determines that for any digital sampling length the bandwidth will always be one-half of the sampling rate, as discussed in Section 6.2. This means that samples taken at a rate of 44 kHz would give 22 kHz (22 000) pictures or snapshots of the waveform in one second. A higher rate will give more samples per second and will also take up more computer memory. The *Nyquist frequency* is the frequency of the highest component of a sound (here 22 kHz) and the *Nyquist rate* is twice the Nyquist frequency.

More computer memory is also used when the bit resolution is higher (16 bits to represent a number versus 8 bits). The available computer memory and the content that is being sampled will determine the sampling rate and bit resolution. For example, sounds that do not have a high-frequency content can be sampled at a lower rate with high fidelity.

It is important that the signal that is being sampled does not have frequencies above the Nyquist frequency. Every time that a sample is made, duplicates of the signal are

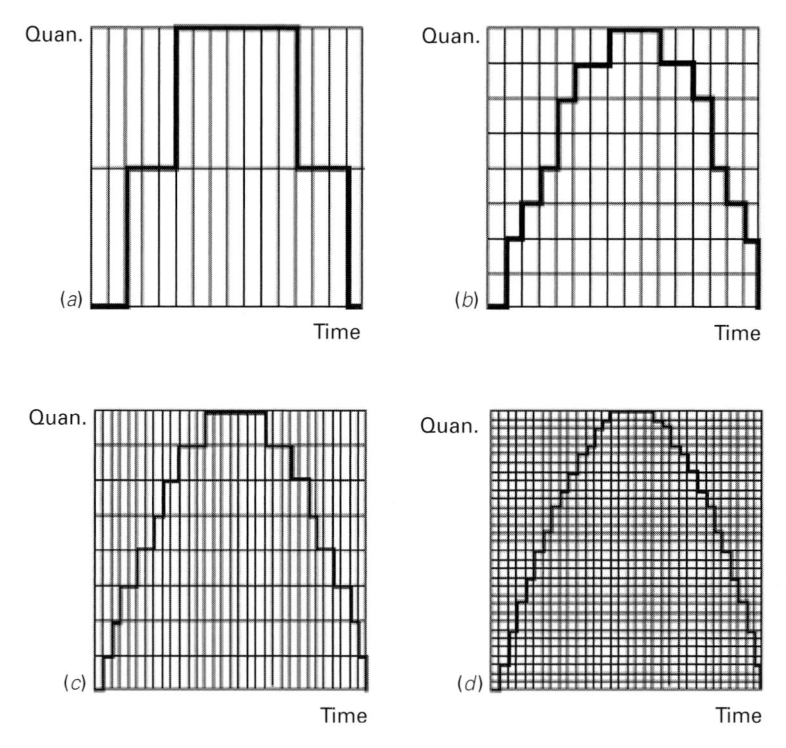

Figure 6.7 Different values of quantization (bit resolution) and sampling rate.

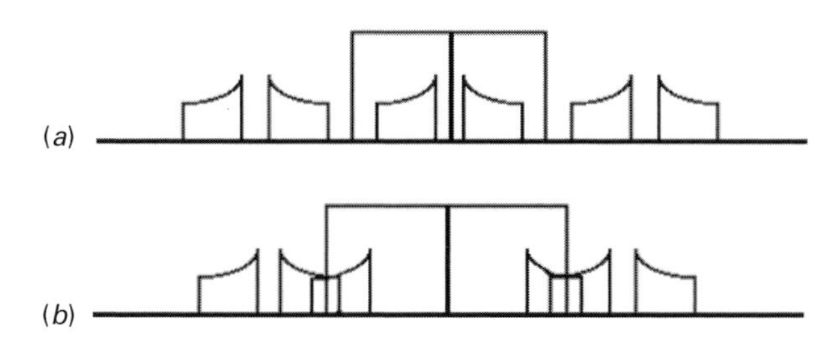

Figure 6.8 Example of (*a*) folding components and (*b*) folding frequency.

also created, called *folding components*. When a sample goes beyond the sampling rate, the duplicate of the signal crosses over into the sampling range and is known as a *folding frequency*; see Figure 6.8. We use the term *aliasing* to describe the folding frequencies, which are doubles of the frequencies. In the sampling process, filters are used before the ADC to make sure that folding frequencies are excluded when the signal is sampled.

In digital sampling, the number of bits per sample N also determines the signal-to-noise ratio, since the SNR depends on how loud the signal is. The *decibel ratio* of the signal is defined as

$$20 \log \frac{2N}{1} \quad \text{or} \quad 20 \log \frac{2N-1}{0.5}$$

The following table shows the decibel ratio for various values of N.

N	dB
2	12
4	24
8	48
12	72
16	96
24	144

Clearly, digital sampling is a very complicated concept and in order to truly capture an acoustic waveform, a sample will need a large amount of computer memory. The following rule-of-thumb can be used to get an idea of how large the file size is for a sampled audio sound:

$$\text{file size} = \text{signal length in seconds} \times \text{number of channels} \times \text{sampling rate} \times \text{bit resolution}/8$$

Example 1 Consider the analogue signal

$$X_a(t) = 3 \cos 100\pi t$$

1. Determine the minimum required sampling rate to avoid aliasing.
2. Suppose that the signal is sampled at the rate $F_s = 200$ Hz. What is the discrete-time signal obtained after sampling?
3. Suppose that the signal is sampled at the rate $F_s = 75$ Hz. What is the discrete-time signal obtained after sampling?
4. What is the frequency $F < F_s/2$ of a sinusoid that yields samples identical to those obtained in part 3?

Solution

1. The frequency of the analogue signal is $F = 50$ Hz. Hence the minimum sampling rate required to avoid aliasing is $F_s = 100$ Hz.
2. If the signal is sampled at $F_s = 200$ Hz, the discrete-time signal is

$$X_n = 3 \cos \frac{100\pi}{200} n = 3 \cos \frac{\pi n}{2}$$

3. If the signal is sampled at $F_s = 75$ Hz, the discrete-time signal is

$$\begin{aligned} X_n = 3\cos\frac{100\pi}{75}n &= 3\cos\frac{4\pi}{3}n \\ &= 3\cos\left(2\pi - \frac{2\pi}{3}\right)n \\ &= 3\cos\frac{2\pi n}{3} \end{aligned}$$

4. For the sampling rate $F_s = 75$ Hz, we have

$$F = fF_s = 75f$$

where f is the frequency of the signal in part 3. Since $f = \frac{1}{3}$, $F = 25$ Hz. Clearly the sinusoid signal

$$\begin{aligned} y_a(t) &= 3\cos 2\pi Ft \\ &= 3\cos 50\pi t \end{aligned}$$

sampled at F_s samples per second yields identical samples. Hence $F = 50$ Hz is an alias of $F = 25$ Hz for the sampling rate $F_s = 75$ Hz.

Example 2 Suppose that we are sampling a four-second speaking-voice sound signal. Because a voice signal is in the lower range of the audio spectrum we could sample the sound at 11 kHz with one mono channel and 8 bit resolution. Using the formula for the required file size we obtain

$$\begin{aligned} \text{file size} &= (4\,\text{s}) \times (1\ \text{channel}) \times (11\ \text{kHz}) \times (8\ \text{bits})/8 \\ &= 44\ \text{kilobytes}\ (44\ \text{k}) \end{aligned}$$

Example 3 Consider a four-second signal comprising a musical selection. First we need to change the sampling rate from its value in Example 2 to 44 kHz, to capture the spectrum of the complex musical sound, and then we will record a stereo sample with 16 bit resolution.

$$\begin{aligned} \text{file size} &= (4\,\text{s}) \times (2\ \text{channels}) \times (44\ \text{kHz}) \times (16\ \text{bits})/8 \\ &= 704\ \text{kilobytes}\ (704\ \text{k}) \end{aligned}$$

The sound signals in Examples 2 and 3 both occupy four seconds, but the final output is very different in size (44 k versus 704 k). If we tried to take Example 2 and record a minutes' worth of sound, the sound file would expand to 10 560 k or 10.6 megabytes, while a minutes' worth of sound from Example 1 would take 660 k. Both examples are one minute long, but the difference in the size of the files is 9900 k or 9.9 M.

6.4 Template-matching techniques

The problem of pattern recognition usually involves the discrimination or classification of a set of processes or events. The number of pattern classes is often determined by the particular application in mind. For example, consider the problem of character recognition for the English language; we have a study of 26 classes. In pattern recognition, comparison is made between a test or input pattern, representing the unknown to be recognized or identified, and one or more *reference patterns*, which characterize known items [5]. Each pattern takes the form of a vector, each element in the vector being the measured value of some feature (a feature is a measurable characteristic of the input that has been found to be useful for recognition). In general the pattern vector takes the form of a set of values for each feature; such a set forms a *template*.

When researchers began using computers to recognize speech, they employed *template matching*. In the template-based approach, each unit of speech (word or phrase) is represented by a template in the same form as the speech input itself. In order for the computer to recognize a particular word or phrase among many, it needs to compare the pattern of the uttered word with all the stored templates in the database.

Matching is a generic operation in pattern recognition; it is used to determine the similarity between two entities (points, curves or shapes). In template matching, a template (typically, a two-dimensional shape) or a prototype of the pattern to be recognized is available. The pattern to be recognized (the input or test pattern) is matched against the stored template while taking into account all allowable pose changes (by translation and/or rotation) and scale changes. The similarity measure, often a correlation, may be optimized from the available training set. Often, the template itself is learned from the training set; in other words, if the input pattern matches the template of the ith pattern class better than it matches any other template then the input is classified as being the ith pattern class [6]. Distance metrics are used to compare templates to find the best match, and dynamic programming is used to resolve the problem of temporal variability.

The classical way of solving a word recognition problem is to treat it as a pattern recognition problem, where digital signal processing techniques can be applied to obtain a pattern for each word [7]. Thus, in general terms, the template matching approach may be interpreted as a special case of the feature extraction approach in which the templates are stored in terms of feature measurements and a special classification criterion (matching) is used for the classifier.

Time alignment

Speech is a time-dependent process. Several utterances of the same word are likely to have different durations, and utterances of the same word with the same duration will differ in the middle, different parts of the words being spoken at different rates.

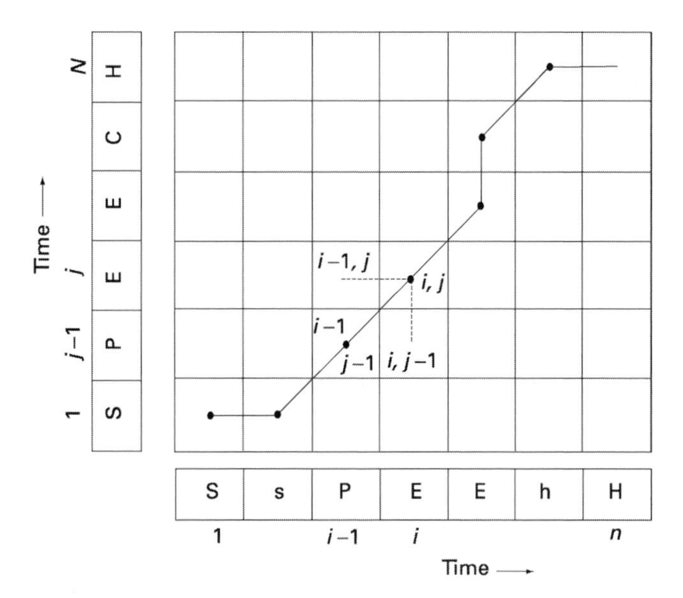

Figure 6.9 Illustration of a time alignment path between a template pattern 'SPEECH' and a noisy input 'SsPEEhH'.

To obtain a global distance between two speech patterns (represented as a sequence of vectors) a time alignment must be performed.

This problem is illustrated in Figure 6.9, in which a time–time matrix is used to visualize the time alignment. As with all time alignment examples the reference pattern (template) is plotted vertically and the input pattern is plotted horizontally. In this illustration the input SsPEEhH is a 'noisy' version of the template SPEECH. The lower-case 'h' indicates a closer match to 'H' than to anything else in the template, similarly for 's'. The input SsPEEhH will be matched against all templates in the system's database. The best matching template is the one for which the path aligning the input pattern and the template is of lowest distance. A simple global distance score for a path is simply the sum of the local distances that go to make up the path.

This algorithm is known as dynamic programming (DP). When applied to template-based speech recognition, it is often referred to as dynamic time warping (DTW). Dynamic programming is guaranteed to find the lowest-distance path through the matrix, while minimizing the amount of computation. The DP algorithm operates in a time-synchronous manner. Each column of the time–time matrix is considered in succession (equivalent to processing the input frame-by-frame) so that, for a template of length N, the maximum number of paths being considered at any time is N.

If $D(i, j)$ is the global distance up to the point (i, j), where i indexes the input-pattern frame and j the template frame, and the local distance at (i, j) is given by $d(i, j)$, how do we find the best-matching path, (i.e. that of lowest global distance) between an input and a template? We could evaluate all possible paths, but this is extremely inefficient as the number of possible paths is exponential in the length of the input. Instead, we will consider what constraints there are on the matching process (or that we can impose on that process) and use those constraints to come up with an efficient algorithm. The constraints we shall impose are straightforward and not very restrictive:

- matching paths cannot go backwards in time;
- every frame in the input must be used in a matching path;
- local distance scores are combined by adding to give a global distance.

For now, it can be said that every frame in the template and the input must be used in a matching path. This means that if we take a point (i, j) in the time–time matrix then the previous point must have been $(i - 1, j - 1)$, $(i - 1, j)$ or $(i, j - 1)$, as shown in Figure 6.9. The key idea in dynamic programming is that at point (i, j) we just continue with the path from $(i - 1, j - 1)$, $(i - 1, j)$ or $(i, j - 1)$ of lowest distance:

$$D(i, j) = \min[D(i - 1, j - 1), D(i - 1, j), D(i, j - 1)] + d(i, j) \tag{6.5}$$

Given that $D(1, 1) = d(1, 1)$ (this is the initial condition), we have the basis for an efficient recursive algorithm for computing $D(i, j)$. The final global distance $D(n, N)$ gives us the overall matching score of the template with the input. The input word is then recognized as the word corresponding to the template with the lowest matching score. (The template length N will be different for each template.)

Basic speech recognition has a small memory requirement; the only storage required by the search (as distinct from the templates) is an array that holds a single column of the time–time matrix. For ease of explanation, it is assumed that the columns and rows are numbered from 0 onwards. This means that the only directions in which the match path can move when at (i, j) in the time–time matrix are already in a form that could be recursively programmed [5]. However, unless the language is optimized for recursion, this method can be slow even for relatively small pattern sizes. Another method, which is both quicker and requires less memory storage, uses two nested 'for' loops. This method only needs two arrays, which hold adjacent columns of the time–time matrix. The path to $(i, 0)$ can only originate from $(i - 1, 0)$. However, the path to a general point (i, j) can originate from the three standard locations (see Figures 6.10 and 6.11). The algorithm to find the lowest global cost is as follows:

1. Calculate column 0 starting at the bottom-most cell. The global cost up to this cell is just its local cost. Then, the global cost for each successive cell is the local cost for that cell plus the global cost to the cell below it. Column 0 is called the predCol (predecessor column).
2. Calculate the global cost up to the bottom-most cell of the next column (the curCol). This the local cost for the cell plus the global cost up to the top-most cell of the previous column.
3. Calculate the global cost of the rest of the cells of curCol. For example, at (i, j) this is the local distance at (i, j) plus the minimum global cost at $(i - 1, j)$, $(i - 1, j - 1)$ or $(i, j - 1)$.

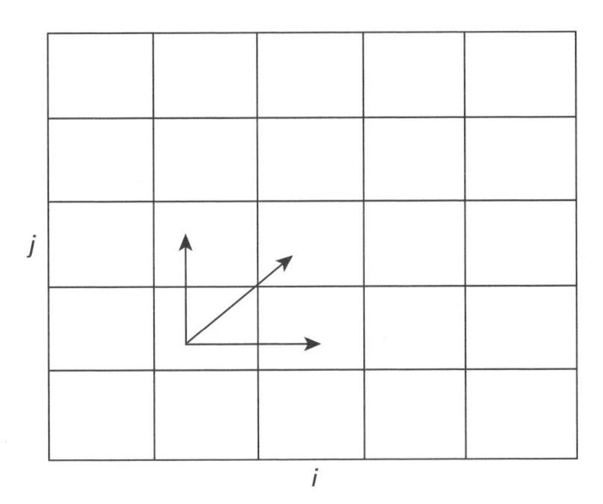

Figure 6.10 The three possible directions in which the best-match path may move from cell (i, j) in symmetric DTW.

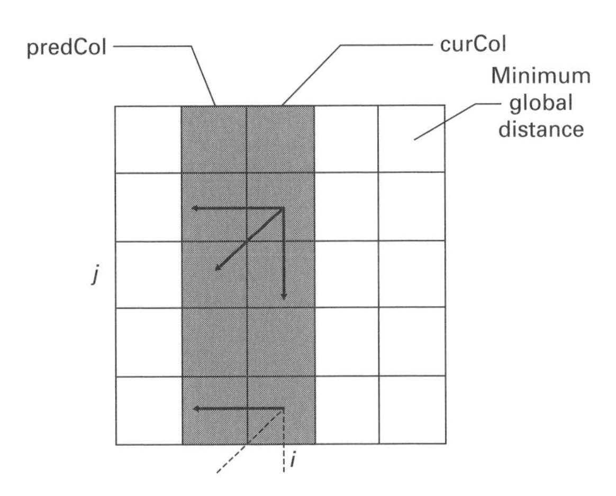

Figure 6.11 The cell at a general point (i, j) has three different possible originator cells. The cell $(i, 0)$ has only one.

4. curCol is assigned to predCol and the procedure is repeated from step 2 until all the columns have been calculated.
5. The global cost is the value stored in the top-most cell of the last column.

6.5 Robustness of recognition systems to natural noise

Speech interference can result in annoyance and tiredness. There are also relationships between individual health and the various effects of noise. Stress can be induced by the presence of noise and can induce physiological changes, which can result in general health decline. One of the most obvious effects of noise on people is its interference with communication. At a fairly well-defined level, noise will mask the sound to be communicated. In speech recognition, performance is rather uniform for SNRs greater than 25 dB, but there is a very steep degradation as the relative noise level increases. It is especially well known that linear prediction analysis, widely used as a front-end speech recognizer, is very sensitive to noise effects. To attenuate problems due to noise distortion, use has been made of noise-cancelling, close-talking and head-mounted microphones, together with speech enhancement and noise-compensation techniques.

Noise induced by the environment or transmission channels can be linear in the power spectrum domain (additive), linear in the log spectral or cepstral domain (convolutional) or simply non-linear in both domains. Obviously the last case is the most difficult to handle. Environment noises are usually additive, but this is not always the case. However, past research concerning speech recognition in noise often considered noise as an additive signal and simulated the noise environment by white-Gaussian or pink noise [9]. Depending on the environment, such assumptions may or may not hold. Another general assumption, often made when dealing with noise, is that the noise is stationary and uncorrelated with the speech signal. The long-term stationarity property of such noise excludes distortions occurring frequently in office environments, such as door slams and speaker-induced noise (e.g. lip-smacks, pops, clicks, coughing, sneezing). Military conditions also provide some examples where techniques developed for stationary noise fail to generalize.

Although much effort has been made to improve the robustness of current recognizers against noise, many laboratory systems or algorithms still assume low noise or model the masking noise with stationary white-Gaussian or pink noise, which does not always represent realistic conditions. Consequently, dramatic degradations have often been observed between laboratory and field conditions.

To address the problem of speech recognition in noisy environments, three approaches have been investigated [1]:

- the improvement of existing systems that have been shown to perform well in laboratory conditions
- the design of new systems that are more robust in noisy conditions
- training the system in an environment similar to that of the testing environment

The third approach is sometimes difficult to apply to practical applications because it is not easy to have identical training and testing environments. A partial solution is to restrict the amount of noise reaching the recognizer, e.g. by using a noise-cancelling microphone. Multi-style training [10] can also help to deal with this problem but further processing is often necessary to achieve good performance.

The techniques that have been developed to recover either the waveform or the parameters of clean speech embedded in noise, and the methods aimed at compensating and adapting the parameters of clean speech before or during the recognition process, depend upon a number of variables. These variables include: the type of noise; the way the noise interacts with the speech signal (which depends upon whether the noise is additive, convolutional, stationary or non-stationary); the number of acquisition channels; the type of communication channel (e.g. a microphone); and the type of *a priori* information about the speech and the noise required by the algorithm. The techniques currently available can be classified in two categories:

- speech enhancement
- model compensation

While speech enhancement focuses on the restoration of a clean speech signal, model compensation allows for the presence of noise in the recognition process itself. Most algorithms use a number of approximations or assumptions. The most common are the following:

- speech and noise are uncorrelated;
- the noise characteristics are fixed during the speech utterance or vary only very slowly (the noise is said to be stationary);
- the noise is supposed to be additive or, in the case of channel distortions, convolutional.

Speech enhancement

Speech enhancement has focused in the past few decades on the suppression of additive background noise. From a signal processing point of view, additive noise is easier to deal with than convolutive noise or non-linear disturbances. Moreover, due to the bursty nature of speech, it is possible to observe the noise by itself during speech pauses, which can be of great value.

Speech enhancement is a very special case of signal estimation, as speech is non-stationary, and the human ear – the final judge – does not find acceptable a simple mathematical error criterion. Therefore subjective measurements of intelligibility and quality are required, and so the goal of speech enhancement is to find an *optimal estimate* $\hat{s}$, i.e., that preferred by a human listener given a noisy measurement $y(t) = s(t) + n(t)$. A number of overview papers can be found in [11, 12].

Speech enhancement by spectral magnitude estimation

The relative unimportance of phase for speech quality has given rise to a family of speech enhancement algorithms based on spectral magnitude estimation (SME). These are frequency-domain estimators in which an estimate of the clean-speech spectral magnitude is recombined with the noisy phase before resynthesis with a standard overlap-add procedure; see Figure 6.12. The name *spectral subtraction* is loosely used for many of the algorithms falling into this class [13] and [14].

Power spectral subtraction

This is the simplest of all the variants. It makes use of the fact that the power spectra of additive independent signals are also additive and that this property is approximately

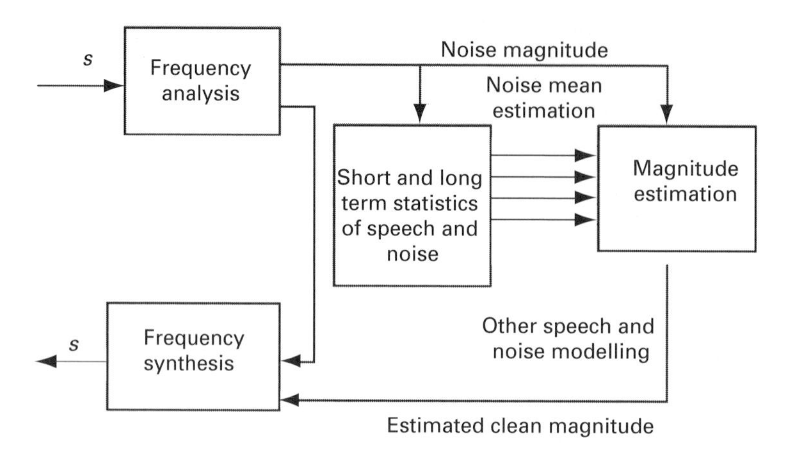

Figure 6.12 Speech enhancement by spectral magnitude estimation.

true for short-time estimates as well. Hence, in the case of stationary noise it suffices to subtract the mean noise power to obtain a least squares estimate of the power spectrum:

$$|\hat{S}(f)|^2 = |Y(f)|^2 - E[|N(f)|^2] \approx |Y(f)|^2 - |N(f)|^2$$

Thus we have

$$\hat{S}(f) = \frac{|\hat{S}(f)|^2}{Y(f)}$$

The greatest asset of spectral subtraction lies in its simplicity and the fact that all that is required is an estimate of the mean noise power and that the algorithm does not need any assumptions concerning the signal. At the same time, the last point is its great weakness. Within the framework occasional negative estimates of the power spectrum can occur. To make the estimates consistent some artificial 'flooring' is required, which yields a very characteristic *musical noise*, caused by the remaining isolated patches of energy in the time–frequency representation.

Much effort has been put into reducing this annoying musical noise. One effective way is the smoothing over time of the short-time spectra. This has the contrary effect, however, of introducing echoes. While reducing the average level of the background noise substantially, plain spectral subtraction has been rather ineffective in improving intelligibility and quality for broadband background noise.

Minimum-mean-square error estimators

Power spectral subtraction is a minimum-mean-square estimator with few or no assumptions about the prior distributions for power spectral values of speech and noise. This is the underlying reason why *ad hoc* operations such as clipping are necessary. Within the framework of spectral magnitude estimation two major improvements are:

- the modelling of realistic *a priori* statistical distributions of speech and noise spectral magnitude coefficients [14]
- minimization of the estimation error in a domain that is perceptually more relevant than the power spectral domain (e.g., the log magnitude domain) [15], [16]

Minimum-mean-square error estimators (MMSEEs) have been developed under various assumptions such as Gaussian sample distributions, a log normal distribution of spectral magnitudes etc. While improving the quality of the output signal, these estimators tend to be complex and computationally demanding.

Time-varying speech models and state-based methods

In the first generation, MMSEEs used a single distribution modelling *all speech* and a single distribution modelling *all noise*. Significant improvement is still possible if one takes into account the non-stationarity of the speech signal (and the noise). The use of *local* speech models implies much smaller variances in the models and tighter estimates. There are two possible approaches:

- the incoming speech is aligned with an ergodic (fully connected) hidden Markov model (HMM) in which a separate MMSEE is associated with each state [17];
- the parameters in a simple parametric speech model are continuously adapted on the basis of the observations [18].

In the first approach a set of possible *states* has to be created during a training phase and this should be a complete set. In the second approach, no explicit training is required but a simpler model may be needed to make continuous parameter updates feasible.

It is obvious that neither state association nor parameter updates will be trivial operations and that this adds another level of complexity to the spectral estimation problem. A side effect of these methods is that they require dynamic time alignment, which is inherently non-causal. While an extra delay of at most a *few frames* is inserted, this may be a concern in some applications.

Wiener filtering

The Wiener filter obtains a least squares estimate of $s(t)$ under stationarity assumptions for speech and noise. The construction of a Wiener filter requires estimates of the power spectrum of the clean speech, $\Phi_{SS}(f)$, and that of the noise, $\Phi_{nn}(f)$

$$W(f) = \frac{\Phi_{SS}(f)}{\Phi_{SS}(f) + \Phi_{nn}(f)}$$

The previous discussion on global and local speech and noise models applies equally to Wiener filtering. Wiener filtering has the disadvantage, however, that the estimation criterion is fixed.

Microphone arrays

Microphone arrays exploit the fact that a speech source is quite stationary and, therefore, by using beam-forming techniques, they can suppress non-stationary interference more effectively than any single-sensor system. The simplest of all approaches is the delay-and-sum beam former, which phase-aligns incoming wave fronts of the desired source before adding them together [19]. This type of processing is robust and needs only limited computational hardware, but it requires a large number of microphones to be effective. An easy way to achieve uniform improvement over the wide bandwidth of speech is to use a subband approach together with a logarithmically spaced array. Different sets of microphones are selected to cover the different frequency ranges [20].

A much more complex alternative is the use of adaptive beam formers, in which case each incoming signal is adaptively filtered before the signals are added together. These arrays are most powerful if the noise source itself is directional. While intrinsically much more powerful than the delay-and-sum beam former, the adaptive beam former is prone to signal distortion in strong reverberation.

A third class of beam formers uses a mix of the previous schemes. A number of digital filters are pre-designed for optimal wideband performance for a set of look directions. An adaptation now exists for selecting the optimal filter at any given moment using a proper tracking mechanism. Under *typical* reverberant conditions, this last approach may prove the best overall solution. It combines the robustness of a simple method with the power of digital filtering. While potentially very powerful, microphone arrays bring about a significant hardware cost owing to the number of microphones and/or adaptive filters required. As a final remark it should be said that, aside from noise suppression, microphone arrays help to de-reverberate the signals as well.

6.6 Application of the correlation dimension

The correlation dimension was originally introduced as a way to measure the dimensionality of a trajectory from a time series of positional data on an orbiting particle in a system [8]. This implies that, when measuring the correlation dimension of the orbit of such a test particle, iteration is required for only that particle and the positional information is recorded at constant time intervals. This by itself presents a computing advantage over the Lyapunov exponent, which requires iteration for both the original particle and a shadow particle.

Before running experiments, it is instructive to have an idea of what results might be theoretically predicted. The question is what type of relationship should be expected between the dimensionality of the orbit and the crossing time. Recall that the definition of dimensionality here basically relates to the manner in which the set fills the space. If an orbit has a high correlation dimension, this implies that it moves very freely through the space, fully exploiting many degrees of freedom. The analysis seems to imply that short-lived orbits will have a larger correlation dimension and that the longest-lived orbits, which are essentially Keplerian, will have rather low correlation dimensions. Therefore, we expect, from this analysis of the problem, that the slope of the relationship between the correlation dimension and the crossing time should be negative.

Before proceeding to discuss the actual findings, let us take a closer look at the method of measuring the correlation dimension. In applications an easily operated definition is

$$C(l) = \lim_{N\to\infty} \frac{1}{N^2} \times \text{ numbers of pairs } (i, j) \text{ whose distance } |\mathbf{X}_i - \mathbf{X}_j| \text{ is less than } l \tag{6.6}$$

This can be calculated by first finding all the distances between every pair (i, j), and then sorting that array. The problem with this method is that it is extremely memory intensive. Since about 10 000 points are required to make an accurate measurement, this results in 5×10^7 pairs of points for which the distance needs to be not only calculated but also stored. Using standard 4 byte floating-point numbers, this would require 200 M of memory.

For that reason, the *binning method* is preferable for this application. Binning received its name because it is very similar to placing items in bins when taking an inventory. The program begins with an array of elements consisting of integers; each element represents a bin. These each begin with the value zero, and each represents a range of distances, whose representative values are from d_{low} to d_{high}. Every time a distance is calculated the appropriate bin is incremented. Note that the lowest bin is incremented for any distance less than d_{low}. When all the distances have been calculated and 'placed' in the appropriate bins, a pass through the array from the second lowest element to the highest is made and the value of each element is then incremented by the value of the one before it. By dividing all the elements by N^2 an approximation to $C(l)$ is created.

From this description, it should be clear that the algorithm used is not overly complex; however, in many respects, neither were the algorithms for measuring the Lyapunov exponent, and they returned data that was impossible to use.

Figure 6.13 shows a typical correlation integral, to which a linear fit can be made to obtain the correlation dimension. There is little question where the curve should be fitted for the measurement, remembering that the short distance scales are important.

Looking at the algorithm in Eqn 6.6, there might be concern about the limit as N goes to infinity. Obviously it will not be possible to use an infinite number of data points, so it is important to know how many are required for the correlation integral to converge. Grassberger and Procaccia [8] found that convergence occurred in their data set rather quickly, within a few thousand data points.

Another factor to consider, when measuring the correlation dimension, that has an impact on the preferred length of integration is the time interval between the storing of data from adjacent positions of the particle. Grassberger and Procaccia [8] urge that

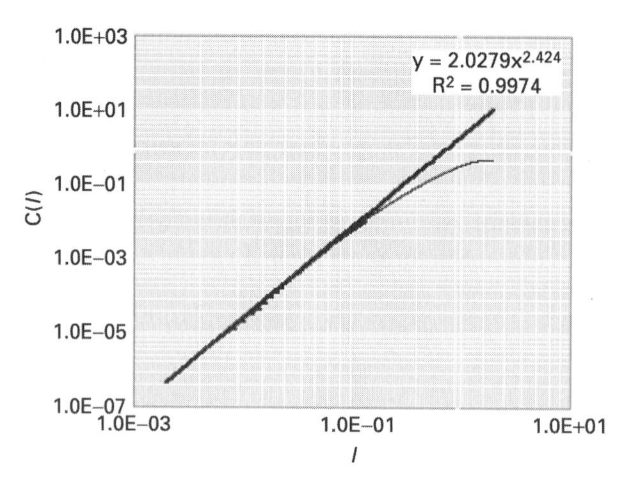

Figure 6.13 The correlation integral for the orbit of a particle between Jupiter and Saturn. To find the correlation dimension $C(l)$, a power-law fit is made to the lower section of this curve. In general, if over 5000 data points are used for calculating the correlation integral, it will always have a slope that is very uniform for the smaller distance scales.

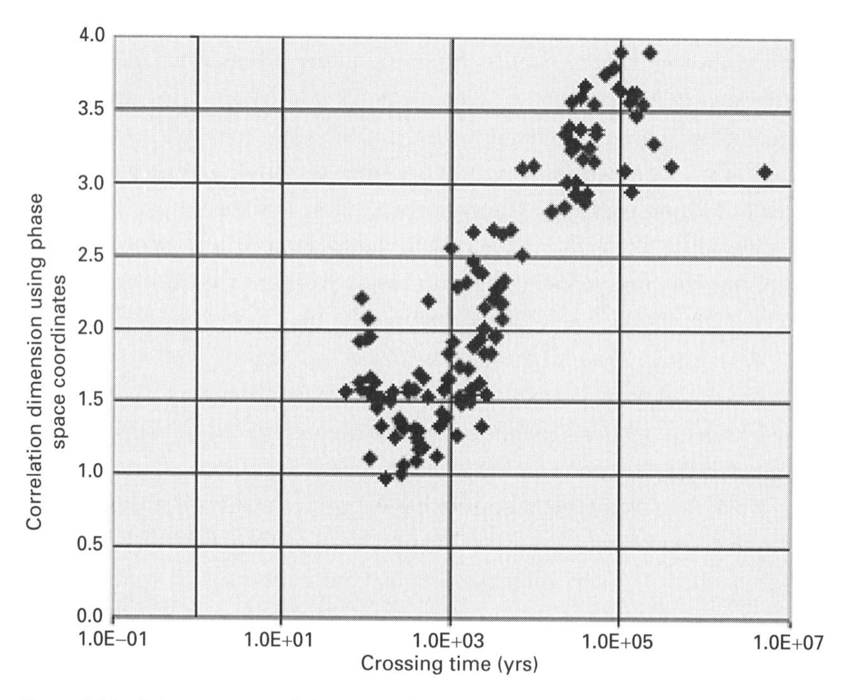

Figure 6.14 Measurements of the correlation dimension versus crossing time for the system of Figure 6.13.

this interval be long enough that adjacent data points are dynamically uncorrelated. Exactly how long this is depends on the exact system. However, this advice is most likely to be of relevance in experimental situations where only one, or possibly two, of the components defining a system can be measured accurately at any one time.

Figure 6.14 shown the measurements of correlation dimension versus crossing time derived from Figure 6.13.

REFERENCES

[1] F. Fallside. *Computer Speech Processing*. Prentice Hall, 1985.
[2] P. Maragos. Measuring the fractal dimension of signals: morphological covers and iterative optimisation. *Proc. IEEE*, **41**(1), 1993.
[3] C. Becchetti and L. Prina Ricotti. *Speech Recognition*. John Wiley and Sons, 1999.
[4] L. R. Rabinar and R. W. Schafer. *Digital Processing of Speech Signals*. Prentice Hall, 1978.
[5] P. Thomas. *Voice and Speech Processing*, pp. 170–96. MacGraw-Hill, 1987.
[6] K. S. Fu. *Digital Pattern Recognition*. Springer-Verlag, Berlin, Heidelberg, second edition, 1980.
[7] M. Al-Zabibi. An acoustic–phonetic approach in automatic Arabic speech recognition. Ph.D. thesis, Loughborough University, Loughborough, UK, 1990.
[8] P. Grassberger and I. Procaccia. Measuring the strangeness of strange attractors. *Physica*, **9D**: 189–208, 1983.
[9] J.-C. Junqua and J.-P. Haton. *Robustness in Automatic Speech Recognition*, pp. 155–82. Kluwer Academic, 1996.
[10] R. Lippmann, E. Martin and D. Paul. Multi-style training for robust isolated-word speech recognition. In *Proc. ICASSP87*, pp. 705–8, 1987.
[11] Y. Ephraim. Statistical-model-based speech enhancement systems. *Proc. IEEE*, **80**(10): 1526–55, 1992.
[12] D. Van Compernolle. Dsp techniques for speech enhancement. In *Proc. ESCA Workshop on Speech Processing in Adverse Conditions*, pp. 21–30, 1992.
[13] M. Berouti, R. Schwartz and J. Makhoul. Enhancement of speech corrupted by additive noise. In *Proc. IEEE Int. Conf. on Acoustics, Speech and Signal Processing*, pp. 208–11, 1979.
[14] Y. Ephraim and D. Malah. Speech enhancement using a minimum mean-square error short-time spectral amplitude estimator. *IEEE Trans. ASSP*, **32**(6): 1109–21, 1984.
[15] R. McAulay and T. Quatieri. Speech analysis/synthesis based on a sinusoidal representation. *IEEE Trans. ASSP*, **34**(4): 744, 1986.
[16] Y. Ephraim and D. Malah. Speech enhancement using a minimum mean-square log-spectral amplitude estimator. *IEEE Trans. ASSP*, **33**(2): 443–5, 1985.
[17] Y. Ephraim, D. Malah and B. H. Juang. Speech enhancement based upon hidden Markov modeling. In *ICASSP 87*, pp. 353–6, 1989.
[18] F. Xie and D. Van Compernolle. Speech enhancement by non-linear spectral estimation – a unifying approach. In *Proc. 3rd European Conf. on Speech Communication and Technology, Berlin*, vol. 1, pp. 617–20, 1993.
[19] J. L. Flanagan. Use of acoustic filtering to control the beamwidth of steered microphone arrays. *J. Acoustical Society of America*, **78**(2): 423–8, 1985.
[20] H. Silverman. Some analysis of microphone arrays for speech data acquisition. *IEEE Trans. Acoustics, Speech and Signal Processing*, **35**(12): 1699–1712, 1987.

7 Speech synthesis using fractals

7.1 Computing fractal noise

Developing mathematical models to simulate and analyse noise has an important role in digital signal and image processing. Computer generated noise is routinely used to test the robustness of different types of algorithm; it is used for data encryption and even to enhance or amplify signals through 'stochastic resonance'. Accurate statistical models for noise (e.g. the probability density function or the characteristic function) are particularly important in image restoration using Bayesian estimation [1], maximum-entropy methods for signal and image reconstruction [2] and in the image segmentation of coherent images in which 'speckle' (arguably a special type of noise, i.e. coherent Gaussian noise) is a prominent feature [3]. The noise characteristics of a given imaging system often dictate the type of filters that are used to process and analyse the data. Noise simulation is also important in the synthesis of images used in computer graphics and computer animation systems, in which fractal noise has a special place (e.g. [4, 5]).

The application of fractal geometry for modelling naturally occurring signals and images is well known. This is due to the fact that the 'statistics' and spectral characteristics of random scaling fractals are consistent with many objects found in nature, a characteristic that is expressed in the term 'statistical self-affinity'. This term refers to random processes whose statistics are scale invariant. An RSF signal is one whose PDF remains the same irrespective of the scale over which the signal is sampled.

Many signals found in nature are statistically self-affine, as discussed in earlier chapters. These include a wide range of noise sources, including background cosmic radiation at most frequencies. In addition, certain speech signals, representative of fricatives, exhibit the characteristics of RSFs, as do other signals such as financial time series, seismic signals and so on. The incredible range of vastly different systems that exhibit random fractal behaviour is leading more and more researchers to consider statistical self-affinity to be a universal law, a law that is particularly evident in systems undergoing a phase transition [6, 7].

Stochastic differential equations of fractional order are considered in subsection 4.1.2 and inverse solutions in subsection 4.3.1, and algorithms are given for the forward and inverse cases.

7.2 Non-stationary algorithms for speech synthesis

Synthesized speech can be produced by several different methods. All these have some benefits and deficiencies, which are discussed in this section. The methods are usually classified into three groups [8–10]:

- articulatory synthesis, which attempts to model the human speech production system directly;
- formant synthesis, which models the pole frequencies of speech signals or transfer functions for the vocal tract using a source-filter model;
- concatenative synthesis, which uses prerecorded samples of different length derived from natural speech.

7.2.1 Formant synthesis

Probably the most widely used synthesis method during the last few decades has been *formant synthesis*, which is based on the source-filter model of speech. There are two basic structures in general, parallel and cascade, but for better performance a combination of these is normally used. Formant synthesis also provides an infinite number of sounds, which makes it more flexible than for example concatenation methods.

At least three formants are generally required to produce intelligible speech and up to five formants to produce high-quality speech. Each formant is usually modelled by a two-pole resonator, which enables both the formant frequency (the pole-pair frequency) and its bandwidth to be specified [11].

Rule-based formant synthesis uses a set of rules to determine the parameters necessary to synthesize a desired utterance [12]. The input parameters may be for example the following, where the open quotient is the ratio of the open-glottis time to the total duration [13]:

- the voicing fundamental frequency (F0)
- the voiced excitation open quotient (OQ)
- the degree of voicing in excitation (VO)
- the formant frequencies and amplitudes (F1–F3 and A1–A3)
- the frequency of an additional low-frequency resonator (FN)
- the intensity of the low- and high-frequency regions (ALF, AHF)

A cascade formant synthesizer, depicted in Figure 7.1, consists of bandpass resonators connected in series; the output of each formant resonator is applied to the input of the following one. The cascade structure needs only the formant frequencies as control

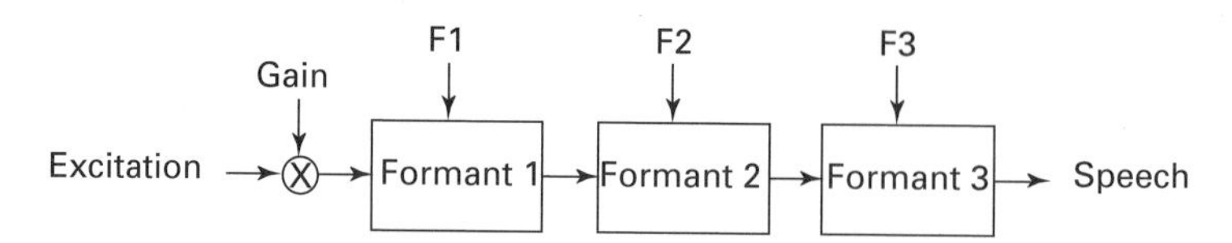

Figure 7.1 Basic structure of cascade formant synthesizer.

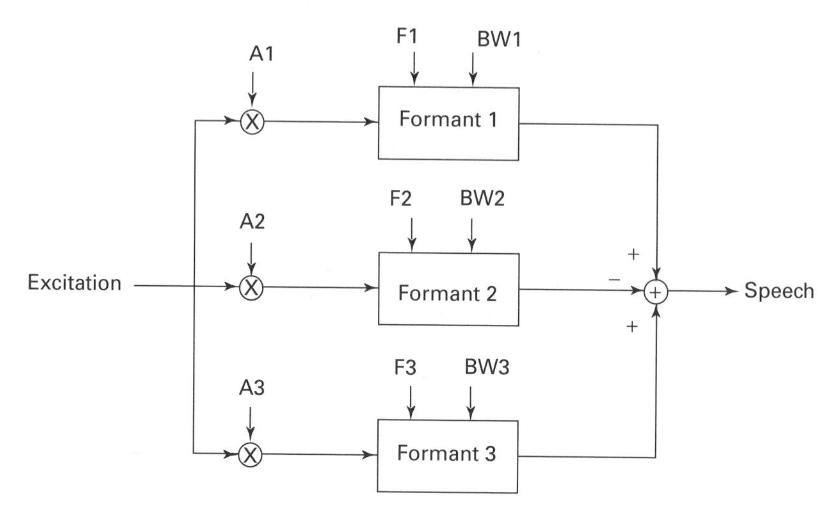

Figure 7.2 Basic structure of a parallel formant synthesizer.

information. The main advantage of the cascade structure is that the relative formant amplitudes for vowels do not need individual controls [12].

The cascade structure has been found better than the parallel structure for non-nasal voiced sounds and because it needs less control information it is simpler to implement. However, with the cascade model the generation of fricatives and plosive bursts is a problem.

A parallel formant synthesizer, Figure 7.2, consists of resonators connected in parallel. Sometimes extra resonators for nasals are used. The excitation signal is applied to all formants simultaneously and their outputs are summed. Adjacent outputs of formant resonators must be summed in opposite phase to avoid unwanted zeros or anti-resonances in the frequency response [10]. The parallel structure enables the controlling of bandwidth and gain for each formant individually and thus needs more control information.

The parallel structure has been found to be better for nasals, fricatives and stop consonants, but some vowels cannot be modelled with a parallel formant synthesizer as well as with a cascade synthesizer.

There has been widespread controversy over the quality of and suitable characteristics for these two structures. It is easy to see that good results are difficult to achieve with only one basic method, so some efforts have been made to improve and combine

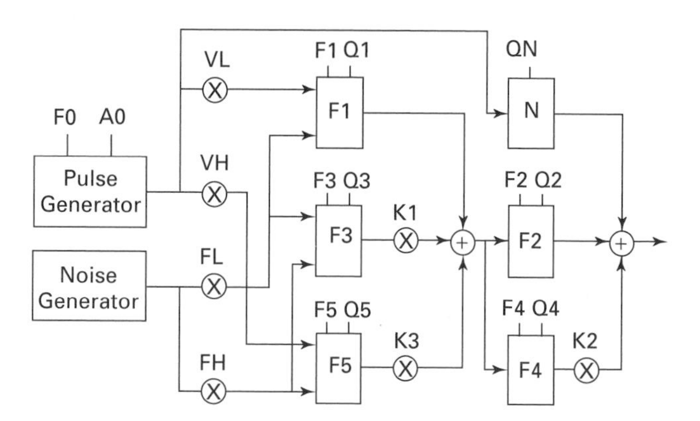

Figure 7.3 The PARCAS model [15].

these basic models. Dennis Klatt [14] proposed a more complex formant synthesizer that incorporated both cascade and parallel synthesizers, with additional resonances and anti-resonances for nasalized sounds, a sixth formant for high-frequency noise, a bypass path to give a flat transfer function, and radiation characteristics. The system used a quite complex excitation model, which was controlled by 39 parameters updated every 5 ms. The quality of the Klatt formant synthesizer was very promising, and the model has been incorporated into several presently operating TTS systems, such as MITalk, DECtalk, Prose-2000 and Klattalk [11].

Parallel and cascade structures can also be combined in several other ways. One solution is to use the so-called PARCAS (parallel cascade) model introduced and patented by Laine [15] in connection with the SYNTE3 speech synthesizer for Finnish. In this model, presented in Figure 7.3, the transfer function for the uniform vocal tract is modelled by two partial transfer functions, each including alternately every second formant of the transfer function. The formant-frequency coefficients $K1$, $K2$ and $K3$ are constant and chosen to balance the formant amplitudes in the neutral vowel in order to keep the gains of parallel branches constant for all sounds. The PARCAS model uses a total of 16 control parameters:

- F0 and A0, the voicing fundamental frequency and amplitude
- Fn and Qn for $n = 1$–5, the formant frequencies and Q-values (formant frequency bandwidths)
- VL and VH, the voiced component amplitudes at low and high frequencies
- FL and FH, the unvoiced component amplitudes at low and high frequencies
- QN, the Q-value of the nasal formant at 250 Hz

The excitation signal used in formant synthesis consists of some kind of voiced source or white noise. The first voiced source signals used were of the simple saw-tooth type: Dennis Klatt [8] introduced a more sophisticated voicing source for his Klattalk system. Correct and carefully selected excitation is important especially when a careful control of speech characteristics is wanted.

The formant filters represent only the resonances of the vocal tract, so additional provision is needed for the effects of the shape of the glottal waveform and the radiation characteristics of the mouth. Usually the glottal waveform is approximated simply by a −12 dB per octave filter and the radiation characteristics by a simple +6 dB per octave filter.

7.2.2 Concatenative synthesis

Connecting prerecorded natural utterances, *concatenative synthesis*, is probably the easiest way to produce intelligible and natural-sounding synthetic speech. However, concatenative synthesizers are usually limited to one speaker and one voice and also usually require more memory capacity than other methods.

One of the most important requirements in concatenative synthesis is to find the correct unit length for the prerecorded utterances. The selection process is usually a trade-off between longer and shorter units. With longer units a high degree of naturalness, fewer concatenation points and good control of co-articulation are achieved, but the amount of units and memory required is increased. With shorter units, less memory is needed but the sample collecting and labelling procedures become more difficult and complex. In the present systems the units used are usually words, syllables, demisyllables, phonemes, diphones and sometimes even triphones (see below).

Single words are perhaps the most natural units for written text and for some messaging systems with very limited vocabulary. The concatenation of words is relatively easy to perform and co-articulation effects within a word are captured in the stored units. However, there is a great difference between words spoken in isolation and those spoken in continuous sentences, which makes synthesized continuous speech sound very unnatural [12].

The number of different *syllables* in each language is considerably smaller than the number of words, but the size of the unit database is usually still too large for TTS systems. For example, there are about 10 000 syllables in English. Unlike with words, the co-articulation effect is not included in the stored units, so using syllables as a basic unit is not very sensible. There is also no way to control the prosodic contours of the sentence.

At the moment, no word- or syllable-based full TTS system exists. The current synthesis systems are mostly based on using phonemes, diphones, demisyllables or some kind of combination of these.

Demisyllables represent the initial and final parts of syllables. One advantage of demisyllables is that only about 1000 of them are needed to construct the 10 000 syllables of English [11]. Using demisyllables instead of, for example, phonemes and diphones requires considerably fewer concatenation points. Demisyllables also take account of most transitions and a large number of co-articulation effects, and they cover a large number of allophonic variations due to the separation of initial and final consonant clusters. The memory requirements are still quite high, but tolerable. In contrast with phonemes and diphones, the exact number of demisyllables in a language cannot be defined. With a purely demisyllable-based system, all possible words cannot be synthesized properly. This problem is faced at the very least with some proper names. However, demisyllables and syllables may be successfully used in a system that uses variable-length units and affixes, such as the HADIFIX system [16, 17].

Phonemes (groups of closely related phones or single sounds) are probably the most commonly used units in speech synthesis because they are the building blocks of the normal linguistic presentation of speech. The inventory of basic phoneme units is usually between 40 and 50, which is clearly much smaller than those for other units [12]. Using phonemes gives maximum flexibility with rule-based systems. However, some phones that do not have a steady-state target position, such as plosives, are difficult to synthesize. The articulation must also be formulated by rules. Phonemes are sometimes used as the input in a speech pre-synthesizer that drives, for example, a diphone-based synthesizer.

Diphones (or *dyads*) are defined to extend from the central point of the steady-state part of a phone to the central point of the following one, so they contain the transitions between adjacent phones. This means that the concatenation point will be in the region of the signal that is most nearly steady state, which reduces the distortion from concatenation points. Another advantage of diphones is that the co-articulation effect no longer needs to be formulated as a set of rules. In principle, the number of diphones is the square of the number of phonemes (plus allophones), but not all combinations of phonemes are needed. For example, in Finnish combinations such as /hs/, /sj/, /mt/, /nk/ or /hp/ within a word are not possible. The number of diphone units is usually from 1500 to 2000, which increases the memory requirements and makes data collection more difficult than for phonemes. However, the database size is still tolerable and, with its other advantages, the diphone is a very suitable unit for sample-based text-to-speech synthesis. The number of diphones may be reduced by inverting symmetric transitions, for example to obtain /as/ from /sa/. Longer segmental units, such as triphones or tetraphones, are only rarely used.

Triphones are like diphones, but contain one phoneme between the steady-state points; therefore a triphone extends from half-phoneme to phoneme to half-phoneme. In other words, a triphone is a phoneme with a specific left and right context. For English, more than 10 000 units are required [18]. Building the unit inventory consists of three main phases [19]. First, the natural speech must be recorded so that all used units (phonemes) within all possible contexts (allophones) are included. After this, the units must be labelled or segmented from spoken speech data and, finally, the most appropriate units must be chosen. Gathering the samples from natural speech is usually very time consuming. However, some of this work may be done automatically by choosing the input text for the analysis phase properly. The implementation of rules to select the correct samples for concatenation must also be done very carefully.

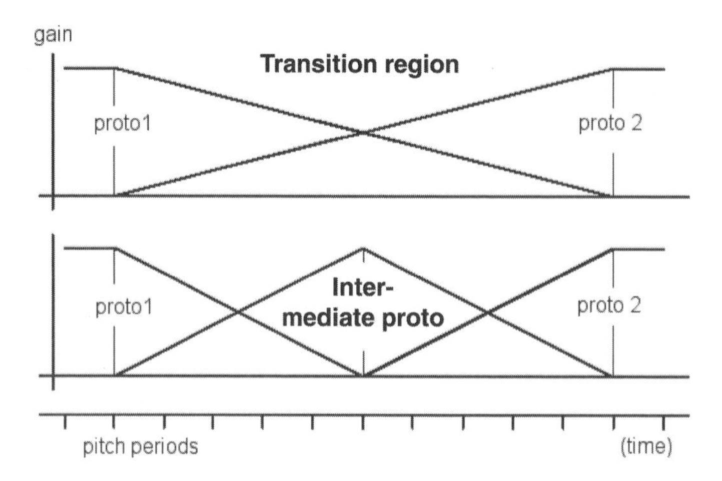

Figure 7.4 Linear amplitude-based interpolation for prototypes [20].

There are several problems in concatenative synthesis compared with other methods:

- distortion from discontinuities at the concatenation points, which can be reduced using diphones or some special methods for smoothing the signal;
- the memory requirements are usually very high, especially when long concatenation units such as syllables or words are used;
- data collection and the labelling of speech samples is usually time consuming. In theory, all possible allophones should be included in the material, but trade-offs between the quality and the number of samples must be made.

Some of these problems may be solved by the methods described below, and the use of the concatenative method is increasing owing to better computer capabilities [11].

The basic idea of the *microphonemic* method is to use variable-length units derived from natural speech [20]. These units may be words, syllables, phonemes (and allophones), pitch periods, transients or noise segments [21]. From these segments a dictionary of *prototypes* is collected. The prototypes are concatenated along the time axis with a PSOLA-like method. If the formant distance between consecutive sound segments is less than two critical bandwidths (barks), concatenation is made by simple linear amplitude-based interpolation between the prototypes. If the difference is more than two barks, an extra, intermediate, prototype must be used because simple amplitude-based interpolation is not sufficient for perceptually acceptable formant movements [20]. The overlap–add processes for prototypes are shown in Figure 7.4.

Some consonants need special attention. For example, stop consonants can be stored using direct waveform segments for different vowel contexts. With fricatives, prototypes of total length about 50 ms and 10 ms units from these are randomly selected for concatenation using the interpolation method described above. Most voiced consonants can be treated like vowels, but the context-dependent variability is higher [21].

The benefits of the microphonemic method are that the computational load and storage requirements are rather low compared with those for other sample-based methods [22]. The biggest problem, as with these other methods, is how to extract the optimal collection of prototypes from natural speech and how to develop rules for concatenating them.

7.2.3 Linear prediction-based methods

Linear predictive (LP) methods were originally designed for speech coding systems but can also be also used in speech synthesis. In fact, the first speech synthesizers were developed from speech coders. Like formant synthesis, the basic LP coder is based on the source-filter model of speech. The digital-filter coefficients are estimated automatically from a frame of natural speech.

The basis of linear prediction is that the current speech sample $y(n)$ can be approximated or predicted from a finite number p of previous samples $y(n-1)$ to $y(n-k)$ by linear combination with a small error term $e(n)$ called the residual signal. Thus

$$y(n) = e(n) + \sum_{k=1}^{p} a(k)y(n-k) \tag{7.1}$$

and

$$e(n) = y(n) - \sum_{k=1}^{p} a(k)y(n-k) = y(n) - \widetilde{y(n)}, \tag{7.2}$$

where $\widetilde{y(n)}$ is the predicted value, p is the linear predictor order and the $a(k)$ are the linear prediction coefficients, which are found by minimizing the sum of the squared errors over the frame. Two methods, the covariance method and the autocorrelation method, are commonly used to calculate these coefficients. Only with the autocorrelation method is the filter guaranteed to be stable [23].

In the synthesis phase, the excitation signal is approximated by a train of impulses for voiced sounds and by random noise for unvoiced sounds. The excitation signal is then gained and filtered with a digital filter whose coefficients are $a(k)$. The filter order is typically between 10 and 12 at an 8 kHz sampling rate, but for higher quality, at a 22 kHz sampling rate, the order needed is between 20 and 24 [24]. The coefficients are usually updated every 5–10 ms.

The main deficiency of the ordinary LP method is that it constitutes an all-pole model, which means that phonemes that contain anti-formants such as nasals and nasalized vowels are poorly modelled. The quality is also poor for short plosives because the time scale for these events may be shorter than the frame size used for analysis. Because of these deficiencies, the speech synthesis quality of the standard LP method is generally considered poor, but with some modifications and extensions to the basic model the quality can be increased.

Warped linear prediction (WLP) takes advantage of human hearing properties, and the needed order of filter is then reduced significantly, from 20–24 to 10–14 with a 22 kHz sampling rate [24, 25]. The basic idea is that the unit delays in digital filters are replaced by the following expression:

$$\tilde{z}^{-1} = D_1(z) = \frac{z^{-1} - \lambda}{1 - \lambda z^{-1}} \tag{7.3}$$

where λ is a warping parameter between -1 and 1 and $D_1(z)$ is a warped delay element; with the bark scale $\lambda = 0.63$ with a sampling rate of 22 kHz. Warped linear prediction provides better frequency resolution at low frequencies and worse at high frequencies. However, this is very similar to human hearing properties [24].

Several other variations of linear prediction have been developed that increase the quality of the basic method [11, 26]. With these methods the excitation signal used is different from that in the ordinary LP method, and the source and filter are no longer separated. These variations on LP include: multipulse linear prediction (MLP), where the complex excitation is constructed from a set of several pulses; residual excited linear prediction (RELP), where the error signal or residual is used as an excitation signal and the speech signal can be reconstructed exactly; and code-excited linear prediction (CELP), where the finite number of excitations used are stored in a finite codebook [27].

7.2.4 Sinusoidal models

Sinusoidal models are based on the well-known assumption that a speech signal can be represented as a sum of sine waves with time-varying amplitudes and frequencies [24, 28, 29]. In the basic model, the speech signal $s(n)$ is modelled as the sum of a small number L of sinusoids:

$$s(n) = \sum_{l=1}^{L} A_l \cos(\omega_l n + \phi_l), \tag{7.4}$$

where $A_l(n)$ and ϕ_l represent the amplitude and phase of each sinusoidal component associated with the frequency track. To find these parameters $A_l(n)$ and ϕ_l, the discrete Fourier transform (DFT) of windowed signal frames is calculated, and the peaks of the spectral magnitude are selected from each frame (Figure 7.5). The basic model is also known as the McAulay–Quatieri model. It has some modified forms, such as the analysis by synthesis / overlap-add (ABS/OLA) and the hybrid / sinusoidal noise models [29].

While sinusoidal models are perhaps very suitable for representing periodic signals, such as vowels and voiced consonants, the representation of unvoiced speech becomes problematic [29].

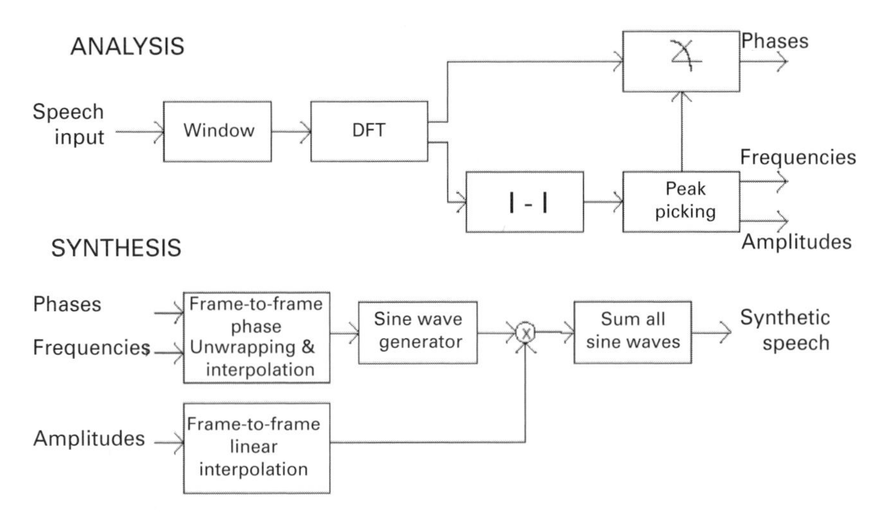

Figure 7.5 Sinusoidal analysis and synthesis systems. [29].

Sinusoidal models are also used successfully in singing-voice synthesis [29, 30]. The synthesis of singing differs from ordinary speech synthesis in many ways. In singing, the intelligibility of the phonemic message is often secondary to the intonation and musical qualities. Vowels are usually sustained longer in singing than in normal speech. An easy and independent controlling of pitch and loudness is required for the synthesis of singing. The best-known system is perhaps the LYRICOS, which was developed at the Georgia Institute of Technology. The system uses sinusoidal-modelled segments from an inventory of singing-voice data collected from human vocalists and maintaining the characteristics and perceived identity. The system uses a standard MIDI interface where the user specifies a musical score, phonetically spelled lyrics and control parameters such as vibrato and vocal effort [30].

7.2.5 Other methods and techniques

Several other methods and experiments to improve the quality of synthetic speech have been made. Variations and combinations of the methods previously described have been studied widely, but there is still no single method to be considered distinctly the best.

To achieve improvements, synthesized speech can also be manipulated afterwards using the normal speech processing algorithms; for example, adding some echo may produce more pleasant speech. However, this approach may easily increase the computational load of the system. Some experiments investigating a combination of the basic synthesis methods have been made, because the different methods show different success in generating individual phonemes. Time domain synthesis can produce high-quality and natural-sounding speech segments, but in some segment combinations the

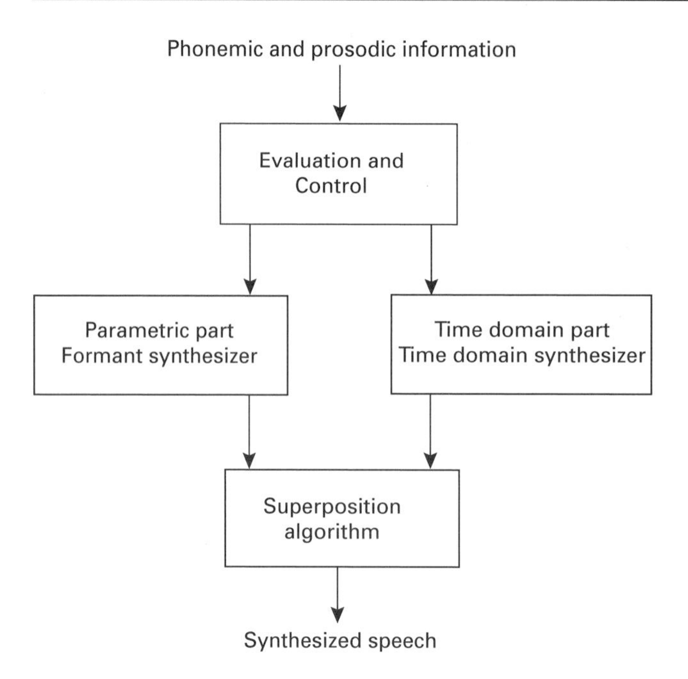

Figure 7.6 Basic idea of the hybrid synthesis system.

synthesized speech is discontinuous at the segment boundaries and, if a wide-range variation in fundamental frequency is required, the overall complexity will increase. Formant synthesis, however, yields more homogeneous speech, allowing a good control of the fundamental frequency, but the voice timbre sounds more synthetic. This approach leads to a hybrid system that combines the time domain and frequency domain methods. The basic idea of the hybrid system is shown in Figure 7.6 [31].

Several methods and techniques for determining the control parameters for a synthesizer may be used. Recently, artificial-intelligence-based methods, such as artificial neural networks (ANN), have been used to control synthesis parameters such as duration, gain and fundamental frequency [32–34]. Neural networks have been applied in speech synthesis for about 10 years; they use a set of processing elements or nodes analogous to neurons in the brain. These processing elements are interconnected in a network that can identify patterns in data as it is exposed to the data. An example of the use of neural networks with WLP-based speech synthesis is given in Figure 7.7. For more detailed discussions of neural networks in speech synthesis, see for example [35–38] and references therein.

Another common method in speech synthesis, and especially in speech recognition and the analysis of prosodic parameters from speech, is the use of hidden Markov models (HMMs). The method uses a statistical approach to simulate real-life stochastic processes [39]. A hidden Markov model is a collection of states connected by transitions. Each transition carries two sets of probabilities: a transition probability, which provides

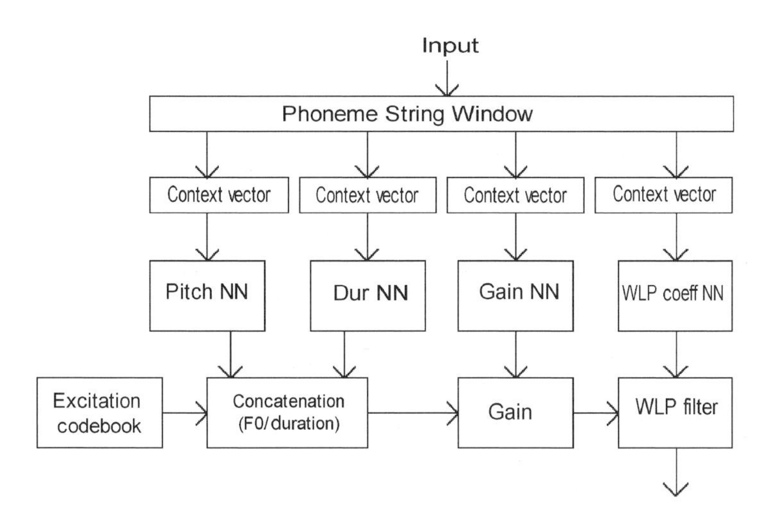

Figure 7.7 An example of the use of neural networks in WLP synthesis [34].

the probability for taking the transition, and an output probability density function, which defines the conditional probability of emitting each output symbol from a finite alphabet given that that the transition has been taken [40].

REFERENCES

[1] B. B. Mandelbrot. *The Fractal Geometry of Nature*. Freeman, 1983.

[2] L. F. Richardson. *The Problem of Contiguity: An Appendix to Statistics of Deadly Quarrels*, vol. 6. General Systems Yearbook, 1961.

[3] A. Rosenfeld and E. B. Troy. Visual texture analysis. Technical Report TR-116, University of Maryland, June 1970.

[4] K. C. Hayes, A. N. Shah and A. Rosenfeld. Texture coarseness: further experiments. *IEEE Trans. Systems, Man and Cybernetics*, **4**: 467–72, 1974.

[5] A. Rosenfeld and A. C. Kak. *Digital Picture Processing*. Academic Press, 1976.

[6] R. Bajcsy. Computer description of textured surfaces. In *Proc. Int. Conf. on Artificial Intelligence*, pp. 572–9, 1973.

[7] R. Bajcsy. Computer identification of textured visual scenes. Technical Report AIM-180, Artificial Intelligence Laboratory, Stanford University, 1972.

[8] D. Klatt. Review of text-to-speech conversion for English. *J. Acoustical Society of America*, **82**(3): 737–93, 1987.

[9] B. Kröger. Minimal rules for articulatory speech synthesis. In *Proc. EUSIPCO92*, vol. 1, pp. 331–4, 1992.

[10] D. O'Shaughnessy. *Speech Communication – Human and Machine*. Addison-Wesley, 1987.

[11] R. Donovan. Trainable speech synthesis. Ph.D. thesis, Cambridge University Engineering Department, England, 1996.

[12] J. Allen, S. Hunnicutt and D. Klatt. *From Text to Speech: The MITalk System*. Cambridge University Press, 1978.

[13] W. Holmes, J. Holmes and M. Judd. Extension of the bandwith of the jsru parallel-formant synthesizer for high quality synthesis of male and female speech. In *Proc. ICASSP90*, vol. 1, pp. 313–16, 1990.

[14] D. Klatt. Software for a cascade/parallel formant synthesizer. *J. Acoustical Society of America* **67**: 971–95, 1980.

[15] U. Laine. Studies on modelling of vocal tract acoustics with applications to speech synthesis. Ph.D. thesis, Helsinki University of Technology, 1989.

[16] H. Dettweiler and W. Hess. Concatenation rules for demisyllable speech synthesis. In *Proc. ICASSP85*, vol. 2, pp. 752–5, 1985.

[17] W. Hess. Speech synthesis – a solved problem? In *Proc. EUSIPCO92*, vol. 1, pp. 37–46, 1992.

[18] X. Huang, A. Acero, H. Hon, Y. Ju, J. Liu, S. Mederith and M. Plumpe. Recent improvements on Microsoft's trainable text-to-speech system – whistler. In *Proc. ICASSP97*, vol. 2, pp. 959–34, 1997.

[19] H. Hon, A. Acero, X. Huang, J. Liu and M. Plumpe. Automatic generation of synthesis units for trainable text-to-speech systems. In *Proc. ICASSP98* (CD-ROM), 1998.

[20] K. Lukaszewicz and M. Karjalainen. Microphonemic method of speech synthesis. In *Proc ICASSP87*, vol. 3, pp. 1426–9, 1987.

[21] L. Lehtinen and M. Karjalainen. Individual sounding speech synthesis by rule using the microphonemic method. In *Proc. Eurospeech*, vol. 2, pp. 180–83, 1989.

[22] L. Lehtinen and A.-A. Puhesynteesi. Speech synthesis in time-domain. Licentiate thesis, University of Helsinki, 1990.

[23] I. Witten. *Principles of Computer Speech*. Academic Press, 1982.

[24] K. Kleijn and K. Paliwal, eds. *Speech Coding and Synthesis*. Elsevier Science, 1998.

[25] U. Laine, M. Karjalainen and T. Altosaar. Warped linear prediction (wlp) in speech synthesis and audio processing. In *proc. ICASSP94*, vol. 3, pp. 349–52, 1994.

[26] D. Childers and H. Hu. Speech synthesis by glottal excited linear prediction. *J. Acoustical Society of America*, **96**(4): 2026–36, 1994.

[27] G. Campos and E. Gouvea. Speech synthesis using the celp algorithm. In *Proc. ICSLP96*, vol. 3, 1996.

[28] R. McAulay and T. Quatieri. Speech analysis/synthesis based on a sinusoidal representation. *IEEE Trans. ASSP*, **34**(4): 744, 1986.

[29] M. Macon and C. Clements. Speech concatenation and synthesis using an overlap-add sinusoidal model. In *Proc. ICASSP96*, pp. 361–4, 1996.

[30] M. Macon, L. Jensen-Link, J. Oliverio, M. Clements and E. George. A singing voice synthesis system based on sinusoidal modeling. In *Proc. ICASSP97*, 1997.

[31] G. Fries. Phoneme-depended speech synthesis in the time and frequency domains. In *Proc. Eurospeech*, vol. 2, pp. 921–4, 1993.

[32] M. Scordilis and J. Gowdy. Neural network based generation of fundamental frequency contours. In *Proc. ICASSP89*, vol. 1, pp. 219–22, 1989.

[33] M. Karjalainen and T. Altosaar. Phoneme duration rules for speech synthesis by neural networks. In *Proc. Eurospeech91*, vol. 2, pp. 633–6, 1991.

[34] M. Karjalainen, T. Altosaar and M. Vainio. Speech synthesis using warped linear prediction and neural networks. In *Proc. ICASSP98*, 1998.

[35] M. Rahim, C. Goodyear, B. Kleijn, J. Schroeter and M. Sondhi. On the use of neural networks in articulatory speech synthesis. *J. Acoustical Society of America*, **2**: 1109–21, 1993.

[36] G. Cawley and B. Noakes. Allophone synthesis using a neural network. In *Proc. First World Congress on Neural Networks*, vol. 2, pp. 122–5, 1993.

[37] G. Cawley and B. Noakes. Lsp speech synthesis using backpropagation networks. In *Proc. IEEE Int. Conf. on Artificial Neural Networks*, pp. 291–3, 1993.

[38] G. Cawley. The application of neural networks to phonetic modelling. Ph.D. thesis, University of Essex, England, 1996.

[39] P. Renzepopoulos and G. Kokkinakis. Multilingual phoneme to grapheme conversion system based on hmm. In *Proc. ICSLP92*, vol. 2, pp. 1191–4, 1992.

[40] K. Lee. Hidden Markov models: past, present, and future. In *Proc. Eurospeech89*, vol 1, pp. 148–55, 1989.

8 Cryptology and chaos

8.1 Introduction

Modern information security manifests itself in many ways, according to the situation and its requirements. It deals with such concepts as confidentiality, data integrity, access control, identification, authentication and authorization. Practical applications, closely related to information security, are private messaging, electronic money, online services and many others.

Cryptography is the study of mathematical techniques related to aspects of information security. The word is derived from the Greek *kryptos*, meaning hidden. Cryptography is closely related to the disciplines of cryptanalysis and cryptology. In simple words, cryptanalysis is the art of breaking cryptosystems, i.e. retrieving the original message without knowing the proper key or forging an electronic signature. Cryptology is the mathematics, such as number theory, and the application of formulas and algorithms that underpin cryptography and cryptanalysis.

Cryptology is a branch of mathematical science describing an ideal world. It is the only instrument that allows the application of strict mathematical methods to design a cryptosystem and estimate its theoretical security. However, real security deals with complex systems involving human beings from the real world. Mathematical strength in a cryptographic algorithm is a necessary but not sufficient requirement for a system to be acceptably secure.

Moreover, in the ideal mathematical world, the cryptographic security of an object can be checked only by means of *proving* its resistance to various kinds of *known* attack. Practical security *does not imply* that the system is secure: other, unknown, types of attack may occur. Nevertheless, practical security against known attacks is certainly the first step in the right direction.

In this chapter we consider the fundamental relationship between two branches of mathematics – cryptology and chaos theory. A brief introduction to each science is presented.

8.2 Cryptology

This section contains a number of basic definitions in cryptology for further reference in this chapter. Comprehensive publications are the excellent books *Handbook of Applied Cryptography* by Menezes *et al.* [1] and *Applied Cryptography* by Schneier [2].

Cryptography is the study of mathematical techniques related to aspects of information security such as confidentiality, data integrity, entity authentication and data origin authentication. One of the goals of this science is keeping messages in electronic communications secret. The message is called the *plaintext*. Encoding the contents of the message in such a way that hides its contents from outsiders is called *encryption*. The encrypted message is called the *ciphertext*. The process of retrieving the plaintext from the ciphertext is called *decryption*. Encryption and decryption usually make use of a *key*, and the coding method is such that decryption can be performed only by knowing the proper key [1].

A method of encryption and decryption is called a *cipher*. Some cryptographic methods rely on the secrecy of the algorithms; such algorithms are only of historical interest and are not adequate for real-world needs. All modern algorithms use a *key* to control encryption and decryption; a message can be decrypted only if the key matches the encryption key.

Cryptanalysis is the study of mathematical techniques for attempting to defeat cryptographic techniques and, more generally, information security services. A goal of cryptanalysis is 'breaking the cipher', i.e. retrieving the plaintext without knowing the proper key.

The following mathematical notation will be used in this chapter,

A A finite set called the *alphabet of definition*.
For digital encryption systems we have a binary alphabet: $A = \{0, 1\}$.

$\prod$ A set called the *plaintext space*.
$\prod$ consists of strings of symbols from an alphabet of definition, i.e. $P \equiv A^*$. An element $p \in P$ is called a *plaintext*.

X A set called the *ciphertext space*.
X consists of strings of symbols from an alphabet of definition C, which may differ from the alphabet of definition for $\prod$. An element $c \in C$ is called a *ciphertext*.

K A set called the *key space*. An element $k \in K$ is called a *key*.

E_e An *encryption function*, which is a bijection from $\prod$ to X uniquely determined by an encryption key $e \in K$, i.e. $E_e : P \rightarrow C$.

D_d A decryption function, which is a bijection from X to $\prod$ using a decryption key $d \in K$; $D_d : C \rightarrow P$.

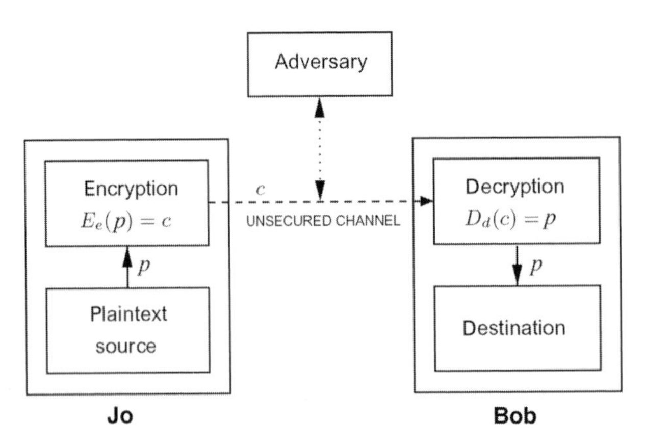

Figure 8.1 Two-party communication using a cipher.

An encryption scheme consists of a set $\{E_e : e \in K\}$ of encryption transformations and a corresponding set $\{D_d : d \in K\}$ of decryption transformations, with the property that for each $e \in K$ there is a unique key $d \in K$ such that $D_d = E_e^{-1}$; that is, $p = D_d(E_e(p))$ for all $p \in P$. An encryption scheme is sometimes referred to as a *cipher* or *cryptographic* system. A simple model of two-party communication using encryption is depicted in Figure 8.1.

Classification

There are two classes of key-based encryption algorithms, *symmetric* (or secret-key) and *asymmetric* (or public-key) algorithms. Symmetric algorithms use the same key for encryption and decryption ($e = d$, or the decryption key is easily derived from the encryption key). Asymmetric algorithms, however, use different keys for encryption and decryption ($e \neq d$), and the decryption key cannot be derived from the encryption key.

Asymmetric ciphers are also called public-key algorithms; their use is referred to in a general sense as public-key cryptography. These ciphers permit the encryption key to be public (it can even be published in a newspaper), allowing anyone to encrypt with the key, whereas only the proper recipient, who knows the decryption key, can decrypt the message. The encryption key is also called the public key and the decryption key the private key or secret key.

Encryption algorithms can be divided into stream ciphers and block ciphers. *Stream ciphers* (also known as *state ciphers*) encrypt a single bit or block of plaintext at a time and the encryption depends on time, i.e. on the cipher state: identical plaintext blocks are transformed into different ciphertext blocks. *Block* ciphers take a larger block of bits and encrypt them as independent units; identical plaintext blocks correspond to identical ciphertext blocks.

Two basic techniques for inserting redundancies into ciphertext should be used, as follows.

Confusion This ensures that the statistical properties of plaintext blocks are not reflected in the corresponding ciphertext blocks. Instead, every ciphertext must have a pseudo-random appearance to any observer or standard statistical test.

Diffusion 1. *In terms of plaintexts*, diffusion demands that (statistically) similar plaintexts result in completely different ciphertexts even when encrypted with the same key. In particular, this requires that any element of the input block influences every element of the output block in a complex irregular fashion.

2. *In terms of the key*, diffusion demands that similar keys result in completely different ciphertexts even when used for encrypting the same block of plaintext. This requires that any element of the pass phrase influences every element of the output block in a complex irregular fashion. Additionally, this property must also be valid for the decryption process because otherwise an intruder might recover parts of the input block from an observed output by a partly correct guess of the pass phrase used for encryption.

8.2.1 Chaos theory

In this subsection we will review the definition of deterministic chaos. The concept and terms can be found in numerous books on chaotic dynamics. A good introductory text is a famous book by Gleick [3] as well as various online resources [4–6]. Fundamentals of the theory and applications are discussed in detail by Cambel [7]. Classic papers are [8, 9].

Formally, *chaos theory* can be defined as the study of complex non-linear dynamic systems. The term '*complex*' implies 'complicated', *non-linear* implies the necessity for recursion and higher mathematical algorithms and *dynamic* implies non-constant and non-periodic. A chaotic system is forever changing, but according to concepts of recursion, either in the form of a recursive process or a set of differential equations modelling a physical or virtual system. Hereafter we discuss *deterministic* chaos only.

8.2.2 Relation between chaos and cryptology

Certain chaotic systems are mathematically deterministic but nearly impossible to predict. Intuitively, this property suits the requirements of a digital encryption system; on the one hand computer-based cryptosystems are deterministic; on the other hand, they must be cryptographically unpredictable. Practically, the last property implies that given certain information on the ciphertext and plaintext, a cryptanalyst should not be able to predict the cryptographic transformation and so recover the whole message.

The well-known cryptographic fundamentals mentioned in the previous subsection, confusion and diffusion, are natural properties of a chaotic system and correspond to topological transivity and sensitivity to initial conditions, as we now describe.

Topological transitivity ensures that any initial state $x_0 \in X$ is transformed to another state $x_n \in X$ in a *key-dependent* and *complex* manner. The transformation $x_0 \xrightarrow{f^n} x_n$ corresponds to the **confusion** requirement of an encryption system. However, the non-linear chaotic map f is not invertible in most cases. To achieve a bijective transformation from the plaintext into the ciphertext, an applicable encryption scheme should be designed.

Sensitivity to initial conditions together with topological transitivity cause the state space X to be 'mixed' by the action of the chaotic map f with the parameters. This corresponds to the **diffusion** requirement in terms of plaintext or key (according to the design of the cryptosystem). In chaos and cryptography we are dealing with systems in which *a small variation in any variable changes the output considerably.*

In both chaos and cryptography, the output – a symbolic trajectory $M_1^l = \{m_1, m_2, \ldots, m_l\}$ in a chaotic system and a pseudo-random sequence or ciphertext in a cryptosystem – *appears* 'random' because of the non-linear transformation of the system state. However, the systems used in chaos are defined on continuous *real numbers*, while traditional cryptography deals with mappings defined on a *finite number of integers*.

8.2.3 Discretization

In order to apply models of continuous chaos to digital cryptography, a valid finite-state approximation or analogue should be used.

If we refer to the properties of the original system, the discretization must satisfy the following characteristic of asymptotic approximation:

$$\lim_{k\to\infty}\max_{x\in X}|f(x) - F(x)| = 0 \tag{8.1}$$

where $f(x)$ is a continuous chaotic function and $F(x)$ is its discretized analogue. In floating-point arithmetic,

$$F(x) = \text{round}_k(f(x)),$$

where $\text{round}_k(z)$ is a function rounding its argument z to precision k.

There are different approaches to *discrete* chaos. A number of theoreticians think that a finite system cannot be viewed as approximately chaotic. We will consider a discrete chaotic system in which the definition of chaos is reduced to the exponential growth of the Hamming distance between states *on a finite interval.*

The two main classes of chaotic discrete-state systems (i.e. the two main approximations) are as follows.

The floating-point approximation to continuous chaos The most obvious approach to the digital implementation of a chaotic system is using floating-point arithmetic (FPA) with a finite precision. The system state x is a vector of floating-point numbers. The map f is evaluated by means of a floating-point CPU. Floating-point arithmetic has been used in numerous cryptographic algorithms [10–14].

Discrete chaos Another approach is to design a discrete-state discrete-time system that exhibits chaotic properties, as defined by Waelbroeck and Zertuche [15]. In practice such a system is based on binary arithmetic (BA). This solution is more natural than the floating-point approximation for a digital device; it is easier to control properties of the system, for instance, cycle lengths. A non-linear binary map, for instance an S-box, is widely used in conventional cryptography. This type of chaos-based encryption resembles ordinary iterated block ciphers or modular random generators (in fact, they all amount to the same thing). Numerous cryptographic techniques rely on discrete chaotic systems such as cellular automata, Kolmogorov flows and chaotic random generators based on bit rotation [16–19].

The main difference between the two approaches is that in the floating-point approximation the metric is difference-evaluated by means of floating-point arithmetic, whereas in discrete chaos the metric is the Hamming distance or equivalent topological metric. On the whole, both approaches use binary arithmetic.

Discrete chaotic cryptology (DCC) is a branch of cryptology in which studies are made of encryption and cryptanalysis based on either discrete approximations of continuous chaos or discrete chaos.

8.2.4 Floating-point approximation to continuous chaos

A floating-point representation of real numbers [20] and a software/hardware implementation of mathematical functions together provide an approximation to a continuous system on a finite-state machine. We can describe a number x in floating-point arithmetic as follows:

$$x = b_m b_{m-1} \ldots b_1 \cdot a_1 a_2 \ldots a_s$$

where a_i, b_j are bits, $b_m b_{m-1} \ldots b_l$ is the exponent and $a_1 a_2 \ldots a_s$ the mantissa of x. Instead of $x_n + 1 = f(x_n)$, where f is the iterated function, we write

$$x_{n+1} = \text{round}_k(f(x_n))$$

where $k \leq s$ and $\text{round}_k(x)$ is a rounding function, here defined as

$$\text{round}_k(x) = b_m b_{m-1} \ldots b_1 \cdot a_1 a_2 \ldots (a_k + a_{k+1})$$

The central problem is that $\text{round}_k(x)$ is applied in each iteration, which leads to 'non-chaotic' behavior of the finite-state system, i.e. it produces very short cycles and trivial patterns. In fact, a floating-point system does not give a correct approximation of continuous chaos. The limit of asymptotic approximation does not converge (the rounding-off error is amplified in each iteration and the original and approximated trajectories diverge exponentially).

Another problem is that most chaotic systems exhibit non-stationary behavior in a bounded space $X \subset R^N$ (often $0 \in X$) only. If x_n gets out of X owing to the rounding operation (e.g. round(x_n) $= 0$), then the consecutive points $x_{n+1}, x_{n+2}, \ldots$ represent a non-chaotic process, i.e. the system produces constant or periodic numbers.

8.2.5 Chaotic cryptography

The chaos-based encryption system is a quite popular topic for research. We now provide few references, as follows.

Ho [21] focused on software implementations of the encryption techniques of Baptista [13] and Wong [14], as well as performance evaluation and cryptanalysis. He suggested using 4-bit blocks to speed up the cryptographic algorithm. However, this increases ciphertext redundancy – four bits of plaintext are encrypted into a 10-bit ciphertext block. Ho explored various chaos-specific cryptography issues – dynamic mapping from plaintext to floating-point states, trajectory distribution and algorithm complexity.

Cappelletti [22] investigated the performance of the hardware implementation (on an FPGA chip) of Scharinger's method [18], based on the highly unstable non-linear dynamics of chaotic Kolmogorov flows.

Machleid [23] discussed several chaos-based cryptographic techniques implemented in multiple floating-point resolutions and with various rounding strategies; he examined randomness properties and defined initial conditions valid for secure encryption.

Summarizing the techniques of chaotic encryption presented above, the following classifications can be considered.

Stream ciphers versus block ciphers

This is a basic classification in conventional cryptology. Typically, chaotic stream ciphers are based on a long trajectory and depend on the system state. Chaotic block ciphers rely on a small number of iterations (16–50) and encrypt blocks independently from each other. For example, Baptista's encryption scheme [13] is a stream cipher, whereas Gutowitz' cellular automata (CA) algorithm [24] is a block cipher.

Floating-point arithmetic (FPA) versus binary integer arithmetic (BA)

This classification emphasizes whether the iterated function is based on floating-point arithmetic. At the final count, FPA relies on BA, but we differentiate these cryptographic systems because of the more complex behavior of FPA implementations. For example, Bianco and Reed [12] use floating-point arithmetic to evaluate the logistic function. Scharinger [18] obtained a chaotic cipher with binary arithmetic.

Plaintext and key in a chaotic system

Here we focus on how the plaintext and the key are introduced into the chaotic system. A traditional approach is to hide the plaintext in the initial conditions $x_0 = p$ and consider the parameters as the key k [12]. Various techniques of seeding the system by means of the secret key have been proposed. Kotulski and Zczepanski [25], for example, suggested dividing the initial-condition vector x_0 into two parts: $x_0 = (p, k)$. In this case, a pair consisting of the plaintext and the key defines a unique trajectory.

Ciphertext in a chaotic system

This classification accents the fact that different system variables can be assigned to the ciphertext. Most chaotic algorithms store the final state x_n as a ciphertext; nevertheless, there are alternative approaches. Baptista [13] suggested a cryptosystem in which the ciphertext is the number of iterations.

Summary of the use of chaos in cryptography

1. No chaotic algorithm has been applied successfully in the *software* industry. The reasons are obvious.
   - Neither the theory nor the practice of chaos-based cryptography has yet been studied enough.
   - Most of the existing chaotic ciphers are vulnerable to a quite simple attack.
   - Floating-point arithmetic requires more hardware resources than binary arithmetic.
   - Traditional cryptography provides the most satisfactory level of theoretical security (unpredictability) and practical security ('resistance').
2. In the last few years the interest in chaos-based cryptography has increased considerably. The explanations are as follows.
   - Cryptography in general has become more important.
   - People want to find new approaches to cryptography in order not to depend on a certain mathematical approach, such as integer factorization or RSA, to which a solution can in principle be found some day.
   - There is the probability of discovering a 'new generation' of ciphers based on the chaos paradigm.

8.3 Unpredictability and randomness

8.3.1 Introduction

Unpredictability, randomness and incompressibility are important characteristics of cryptographic sequences. Modern ciphers use cryptographic sequences that typically define substitutions and permutations. The complexity of a cryptographic sequence, i.e. its 'degree' of randomness and unpredictability, determines the strength of the encryption algorithm. Limitation of computational resources is the main obstacle that prevents intruders from predicting the key or the plaintext.

The three concepts – unpredictability, randomness and incompressibility – are interlocked with each other, i.e. if a sequence is cryptographically random (i.e. it has a uniform probability distribution) it is also incompressible and unpredictable.

In practice we speak about *computational* unpredictability, randomness and incompressibility, assuming that prediction, observation and compression respectively are carried out by a digital computer. A stream of data or *string* S is said to be *computationally incompressible* or *algorithmically random* if its length equals the length of the shortest program producing it. We say a string is *unpredictable* when there is no efficient instrument that predicts the next bit of the string given all the previous elements. Here, 'predicts' means that the probability of a successful forecast is greater than 1/2 .

The following subsections provide a formal definitions of these terms, based on [26], as well as the necessary background from complexity theory, probability theory [27] and information theory [38]. Another objective is how to decide whether chaotic systems are cryptographically unpredictable and how to find a 'measure' of their randomness.

8.3.2 Probability ensemble

In cryptography, we define a probability distribution function (PDF) as a function p from strings to non-negative real numbers $\pi : \{0, 1\}^* \to [0, 1]$ such that $\sum_{\alpha\in\{0,1\}^*} \pi(\alpha) = 1$. The number $\pi(\alpha)$ is interpreted as the probability of occurrence of the string α.

Two PDFs π_1 and π_2 are equal if they assign identical probability to the same string, i.e.

$$\forall \alpha \in \{0, 1\}^*, \quad \pi_1(\alpha) = \pi_2(\alpha)$$

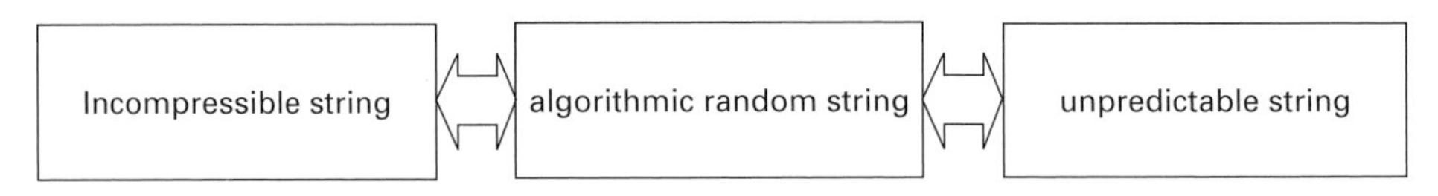

Figure 8.2 Characteristics of cryptographic sequences.

A uniform probability distribution function (UPDF) on the set $S = \{0, 1\}^N$ is a PDF π_0^N such as

$$\forall \alpha, \beta \in S, \quad \pi_0^N(\alpha) = \pi_0^N(\beta)$$

A probability ensemble indexed by I is a sequence $\prod = \pi_{i\,i \in I}$ such that π_i is a probability distribution on some finite domain.

A cumulative distribution function (CDF) is a function ρ from strings to non-negative real numbers $\rho : \{0, 1\}^* \to [0, 1]$ defined as

$$\rho(\alpha) = \sum_{i \in I | \alpha_i < \alpha} \pi(\alpha_i)$$

8.3.3 Algorithmic complexity

We think of a complex string S as being produced by a complex algorithm. High complexity implies difficulty in analysis and prediction, and this is usually needed in cryptography.

The algorithmic complexity $K_U(S)$ of a finite string S with respect to a *universal* Turing machine U is the length $l(\pi)$ of the smallest computer program π that computes it:

$$K_U(S) = \min_{\pi, U(\pi)=S} (l(\pi))$$

where the minimum is taken over all programs π that yield S and halt when processed by U. The complexity with respect to any other Turing machine A is related to $K_U(S)$ via

$$K_U(S) \leq K_A(A) + C_A$$

where C_A is a constant that depends on U and A but not on S.

In practice, it is impossible to find exactly the 'smallest computer program', especially when the algorithm is not trivial. However, we can use the definition of $K_U(S)$ to draw a qualitative comparison between algorithms, in particular between traditional and chaotic pseudo-random sequences.

8.3.4 Complexity

The main goal of complexity theory is to provide mechanisms for classifying computational problems and measuring the amount of resources needed to solve them. The resources measured may include time, storage space, number of processors etc., but typically the main focus is on time and sometimes space. In this subsection, we consider three classes of complexity of computational problems, defined by formal models based on Turing machines [28]:

- polynomial-time computations (P)
- non-deterministic polynomial-time computations (NP)
- bounded-away-error probabilistic polynomial-time computations (BPP)

It is generally believed that P $\neq$ NP and, furthermore, that NP $\neq$ BPP. The complexity classes NP and BPP both consider the worst-case complexity of problems, not the average one. Often we need a stronger assumption than NP $\neq$ BPP, since many problems involve both the second and third categories of complexity [26].

8.3.5 Pseudo-random number generator

Let $G = G_n, n \geq 1$, be an ensemble of generators, with $G_n : 0, 1^n \to 0, 1^{p(n)}$, where $p(\cdot)$ is a polynomial satisfying $n + 1 \leq p(n) \leq n^c + c$ for some fixed integer c. We say that Γ is a pseudo-random-number generator (PRNG) if:

- there is a deterministic polynomial-time Turing machine that on input of any n-bit strings outputs a string of length $p(n)$;
- two probability ensembles $\prod_1 = G_n(\prod_0^n)$ and $\prod_0^{p(n)}$ are polynomially indistinguishable for sufficiently large n.

Let $f : 0, 1^n \to 0, 1^n$ be a one-way function with a hard-core predicate $b : 0, 1^* \to 0, 1$. Let G_n be a polynomial-time Turing machine defined in the following way: for an input x, the machine G_n computes the bits $b_i = b(f^i x))$, where $1 \leq i \leq 2n$. Iteratively applied i times, the function f output the bits $b_{2l(x)} b_{2l(x)-1} \ldots b_1$. Then G_n is unpredictable.

8.3.6 Algorithmic randomness and chaos

We define the complexity K of the trajectory of a point x_1 with respect to an open cover $\beta = X_1, X_2, \ldots, X_m$ of X as

$$K(x_1, f|\beta) = \lim_{n\to\infty} \sup \frac{1}{n} \min_{M_l^n \in [\psi(x_1)]^{-n}} K(M_1^n)$$

where

$$[\psi(x_1)]^n = \{M_1^n | f^{j-1}(x_1) \in X_{m_1}, \quad m_j \in M_l^n, \quad j = 1, 2, \ldots, n\}$$

The complexity of the trajectory of a point x_1 is defined as

$$(x_1, f) = \sup_{\beta} K(x_1, f|\beta)$$

The trajectory of the point x_1 is called *algorithmic random* if its complexity is positive. The trajectories of almost all state points $x \in X$ are algorithmic random with complexity equal to the Kolmogorov–Sinai (KS) entropy of M. It is clear that a PRNG can be constructed on a chaotic system with *positive* KS entropy. However, such a PRNG cannot be implemented on a computer because it has a finite state space.

8.4 Chaotic systems

This section provides an overview of continuous and discrete-state chaotic systems that have already been applied to cryptography. A chaotic system is defined by an iterated function (a chaotic map) f and a state space X. The iterated function f transforms the current state of the system to the next one:

$$x_{n+1} = f(x_n)$$

where $x_n \in X$ denotes the system state at the discrete time n. Often in chaos-based cryptography the state space is the finite binary space

$$X = P = C = 0, 1^n, \qquad n = 1, 2, \ldots$$

The initial conditions form a vector $x_0 \in X$, and it is assigned to an internal state variable before the first iteration. The vector $c \in C = \{0, 1^n\}$ contains the parameters of the dynamic system. The parameters are kept constant through all cycles (iterations). A chaotic sequence is the output of the generator and represents the variation in the state of the chaotic system in time:

$$\text{output} = x_0, x_1, x_2, \ldots, x_n, \ldots$$

The central requirement of a cryptographic sequence is that it should be unpredictable for an external observer. We assume the outsider does not possess critical data about the generator, called the *seed*. In most chaos-based ciphers the seed (or part of it) becomes a key. The seed can consist of the initial conditions, the parameters, the number of iterations etc. The goal is to find a chaotic system such that the observer, even knowing the algorithm and having to hand a reasonably long output sequence:

- can only guess the true seed as being one of a number of equally distributed seeds;
- can only guess the next bit (block) of the cryptographic sequence as one of a number of bits with equally distributed zeros and ones (binary sequences).

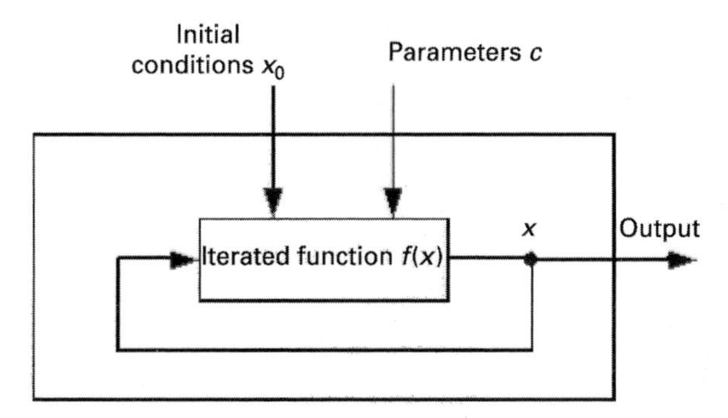

Figure 8.3 Chaos generator.

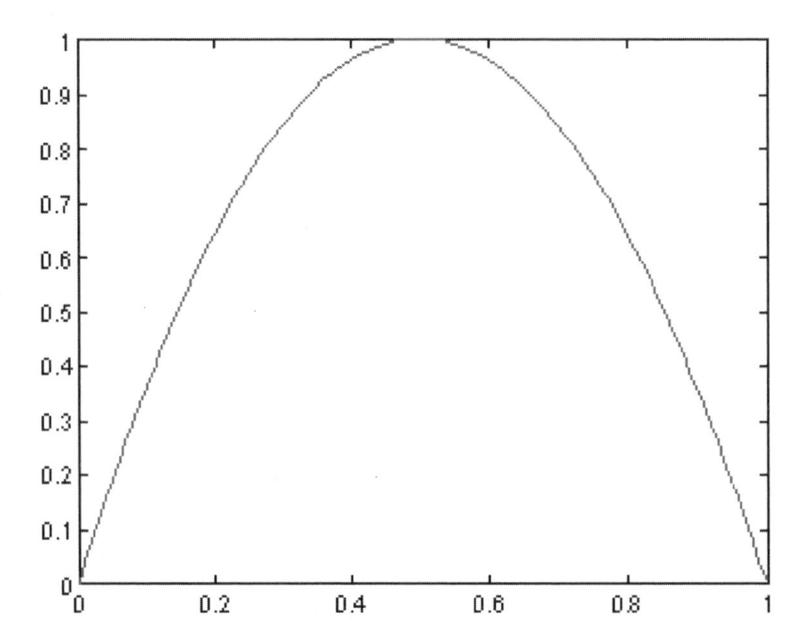

Figure 8.4 The logistic map $f(x)$ at $r = 0.99$.

The probability of recovering the seed and/or predicting the next bit should be sufficiently low to consider the system secure.

The choice of chaotic map depends on the cryptographic application. We now consider various one-dimensional maps in the rest of this subsection, and in the next subsection we will consider two-dimensional maps.

One-dimensional chaotic maps

Logistic map Feigenbaum studied a simple equation, called the logistic map (see Figure 8.4):

$$f(x) = 4 \cdot r \cdot x \cdot (1 - x)$$

where $x \in X = (0, 1)$, $r \in (0, 1)$; X is the ciphertext space, defined at the start of section 8.2. The corresponding sequence of chaotic numbers x_1, x_2, K, x_n, K, where K is the key space, is defined by induction:

$$x_0 \in X, \qquad x_{n+1} = 4 \cdot r \cdot x_n(1 - x_n)$$

A chaotic generator based on a floating-point evaluation of this equation can be easily implemented in a software encryption system. With certain values of the parameter r the generator spits out a sequence of 'pseudo-chaotic' numbers, which appear random, as shown in Figure 8.5. The Freigenbaum cascade diagram shows the values of x on the attractor for each value of the parameter r, as depicted in Figure 8.6. As r increases, the number of points in the attractor increases from 1 to 2 to 4 to 8 and so on to infinity. In this area (r®1) there is no practical way to estimate the final state of the system

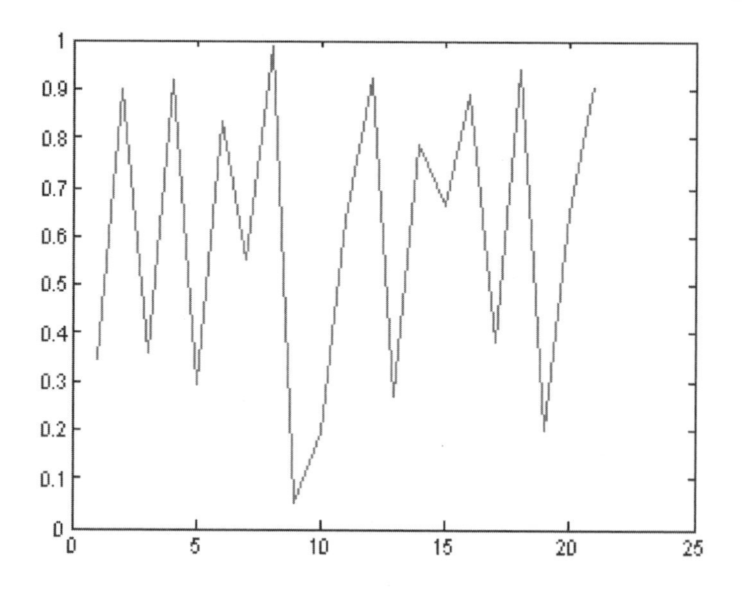

Figure 8.5 Sequence x_1, x_2, K, x_n, K generated using the logistic map at $x_0 = 0.34$, $r = 0.99$.

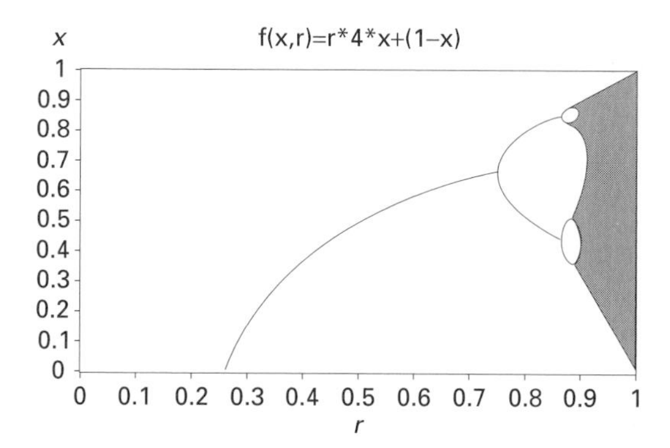

Figure 8.6 Bifurcation of the logistic map.

given the initial conditions x_0 (without performing the n iterations), or, vice versa, to recover x_0, which can be a key or a plaintext, from x_n. This property can be considered as a fundamental advantage of using chaos in cryptography. However, the state can be predicted probabilistically if the state distribution is not uniform. A logistic map produces sequences with a particular PDF and thus is predictable.

For any long sequence of N numbers generated from the logistic map we can calculate the Lyapunov exponent from

$$\lambda(x_0) = \frac{1}{N} \sum_{n=1}^{N} \log |r(1 - 2x_n)|$$

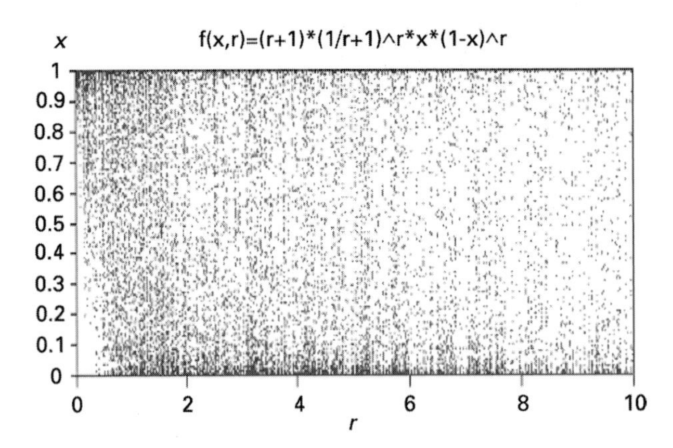

Figure 8.7 Attractor points corresponding to different values of the parameter r in the Matthews map.

Numerical estimation for $r = 0.9$ and $N = 4000$ gives

$$\lambda(0.5) \approx 0.7095.$$

In 1989 Matthews [10] generalized the logistic map with cryptographic constraints and developed the following:

$$f(x) = (r+1) \cdot \left(\frac{1}{r} + 1\right)^r \cdot x \cdot (1-x)^r, \qquad r \in (1, 4)$$

The advantage is that a wide range of the parameter r produces a fractal attractor with lots of points, as shown in Figure 8.7. As a result the parameter r can be used as a secret key that will put the system into an unpredictable state after several cycles.

Tent map A one-dimensional (skew) tent map is defined as

$$f(x) = \begin{cases} ax, & 0 \leq x \leq a \\ \dfrac{1-x}{1-a}, & a < x \leq 1 \end{cases} \tag{8.3}$$

where the parameter $a \in (0, 1)$ determines the top of the tent (see Figure 8.8). The Lyapunov exponent depends on the parameter a:

$$\lambda(a) = -a \ln a - (1-a) \ln(1-a)$$

for almost all $x_0 \in (0, 1)$ [29]. Numerically, $\max_{a \in (0,1)} \lambda(a) \approx 0.693$ (Figure 8.9).

Habutsu *et al.* [11] suggested a cryptosystem for iterating the inverse tent map. The parameters are the key $a \in [0.4, 0.6]$ and a pseudo-random binary sequence $r = r_{i\,i \in I}$.

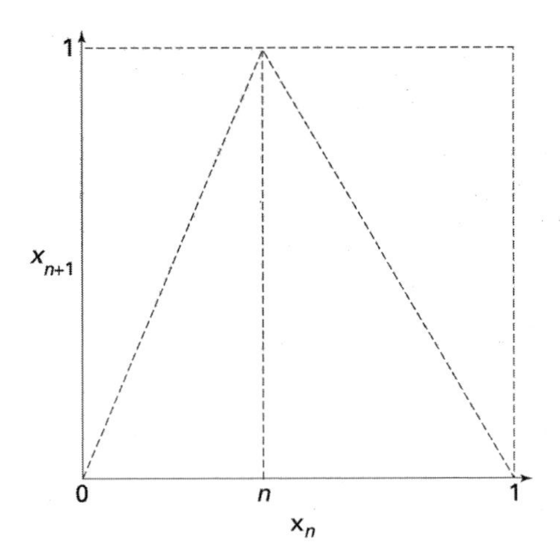

Figure 8.8 The tent map.

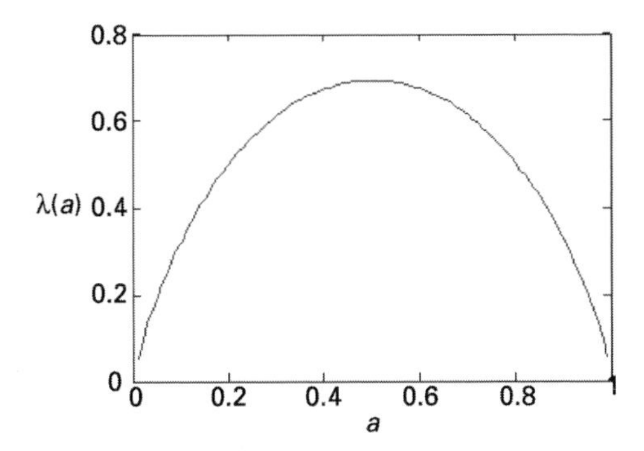

Figure 8.9 Lyapunov Exponent of the tent map as a function of a.

The map can be written as

$$x_{i+1} = \begin{cases} ax_i, \\ (a-1)x_i + 1, \end{cases} \qquad \text{if } r_i = 0, \qquad i \in I \tag{8.4}$$

Masuda and Aihara [30] proposed a discretized version of the tent map (see Figure 8.10):

$$F(X) = \begin{cases} \left\lceil \frac{M}{A} X \right\rceil, & 1 \leq X \leq A \\ \left\lfloor \frac{M}{M-A}(M-X) + 1 \right\rfloor, & A < X \leq M \end{cases} \tag{8.5}$$

where $X, A = 1, 2, \ldots M$ and $\lfloor\,\rfloor$, $\lceil\,\rceil$ indicate the instructions 'round down' and 'round up' respectively.

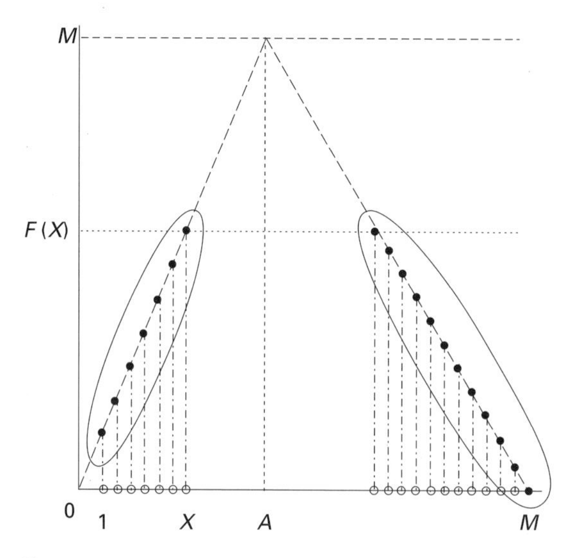

Figure 8.10 The discretized tent map. The large loops show the points that are included.

Chebyshev mixing transformation Erber *et al.* [31] suggested using a Chebyshev mixing polynomial to simulate a random process on digital computers. The iterated function is given by

$$x_{n+1} = x_n^2 - 2, \qquad x_n \in (-2, 2), \qquad x_n \neq -1, 0, 1$$

The output is obtained after an additional non-linear transformation of x_n:

$$y_n = \frac{4}{\pi} \arccos\left(\frac{x_n}{2}\right) - 2$$

Several investigators have found that the Cebyshev transformation possesses undesirable qualities, which make it unsuitable for pseudo-random number generation.

Sawtooth map Consider the map shown in Figure 8.11,

$$f(x) = r \cdot x \bmod q$$

and a sequence x_1, x_2, K, x_n, K such that

$$\ldots_{1} = r \cdot x_n \bmod q$$

ere $n = 0, 1, x_0 \in [0, q], \quad r = p/q > 1$ and p is a co-prime to q. The map is chaotic ll r and has Lyapunov exponent $\lambda = \log r > 0$ [32, 33]. One can consider the linear uential generator (LCG) as a generalized discrete version of the sawtooth map:

$$\ldots (Ax_n + C) \bmod M$$

C and M are fixed by the designer [34].

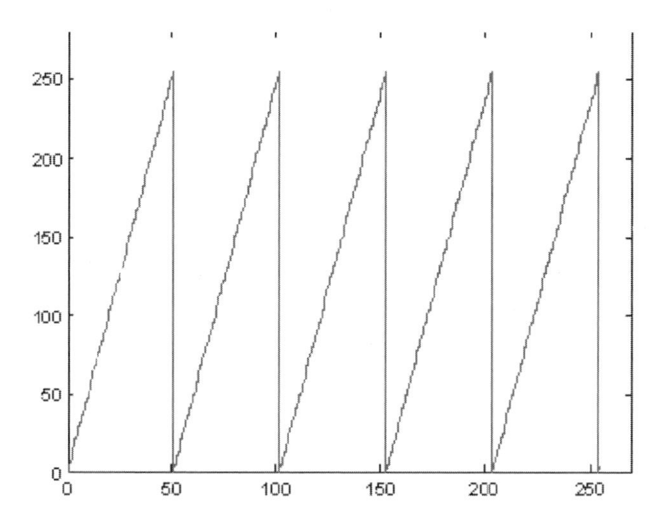

Figure 8.11 Sawtooth map with $p = 1279$, $q = 255$.

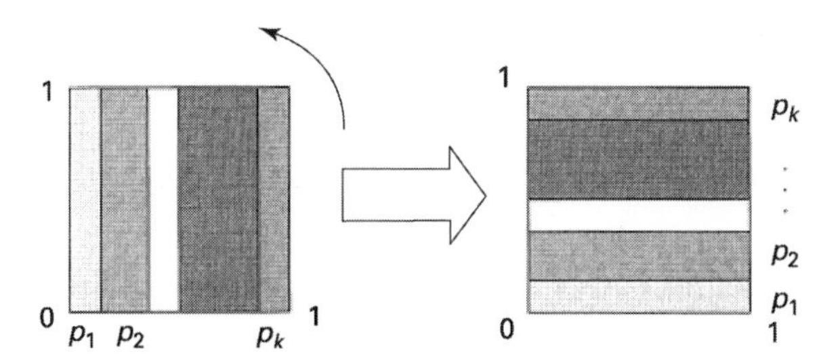

Figure 8.12 T_k mapping.

8.4.1 Two-dimensional chaotic maps

Kolmogorov flows represent the most unstable class of chaotic systems [35] and can be particularly useful for mixing two-dimensional data blocks. The iterated function T_k, also known as a generalized Baker map, can be considered as a geometrical transformation of a square image: the image is divided into vertical strips according to a partition set $p_1, p_2, \ldots, p_k$ and transposed to horizontal strips counterclockwise, as shown in Figure 8.12.

Formally, the continuous Kolmogorov flow $\langle T_k, E \rangle$ can be described as follows. The unit square $E = [0, 1) \times [0, 1)$ is the state space. Let a sequence $\pi \, {p_i}_{i=1}^{k}$, $p_i \in (0, 1)$, $k > 1$, denote a partition of E into vertical strips (Figure 8.12) such as $\sum_{1<i<k} p_i = 1$, and let F_s denote the left-hand borders of these strips:

$$F_i = \begin{cases} 0, & \text{if} \quad i = 1 \\ p_1 + p_2 + \cdots + p_{i-1}, & \text{if} \quad i > 1 \end{cases} \tag{8.6}$$

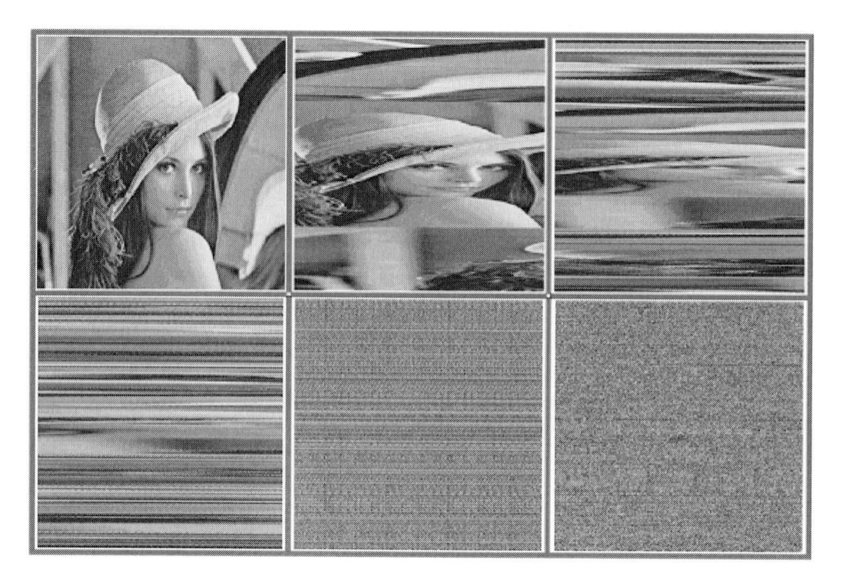

Figure 8.13 Key-dependent permutation with Kolmogorov flow at $\pi(0.25, 0.5, 0.25)^{10}$.

The iterated function $T_k : E \to E$ is given by

$$T_k(x, y) = \left(\frac{1}{\pi}(x - F_i), p_i y + F_i \right)$$

for

$$(x, y) = [F_i, F_i + p_i) \times [0, 1)$$

The transformation T_k stretches each rectangle $[F_{i-1}, F_i) \times [0, 1)$, $i = 1, 2, \ldots, k$ horizontally by a factor $1/p_i$.

The system exhibits a high sensitivity to the initial condition

$$x_0 \in [0, 1) \times [0, 1)$$

and the parameter $\pi = p_{i\,i=1}^{\,k}$. Figure 8.13 illustrates the mixing property: the picture appears equally diffused through the state space after six rounds only.

Since an image is defined on a lattice of finite set of points (pixels), a corresponding discretized form of the continuous map needs to be derived. Fridrich [36] describes a discretized Kolmogorov flow map as follows. Let $T_p : B \to B$, $B = [0, 1^N) \times [0, 1^N)$ denote the iterated function and $\rho = n_1, n_2, \ldots, n_k$ be chosen in such a way that each integer n_i divides N and $n_1 + n_2 + \cdots + n_k = N$. The bit $b_{r,s} \in B$ is mapped to

$$B_p(r, s) = \left(\frac{N}{n_i}(r - N_i) + s \bmod \frac{N}{n_i}, \ \frac{n_i}{N}\left(s - s \bmod \frac{N}{n_i} \right) + N_i \right)$$

where $r, s \in \{1, 2, \ldots, N\}$; $N_i = n_1 + n_2 + \cdots + n_i$, $i \in \{1, 2, \ldots, k\}$, and $N_i \leq r < N_i + n_i$.

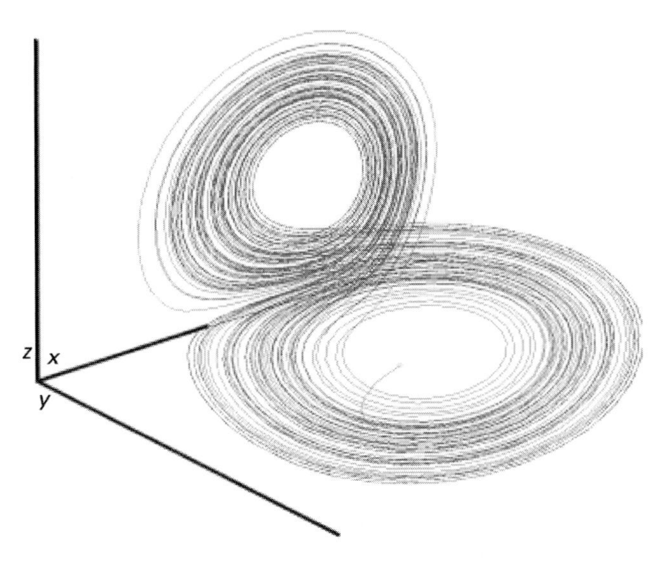

Figure 8.14 Lorenz attractor for parameter values $\rho = 25$, $\sigma = 10$, $\beta = 8/3$.

8.4.2 Multidimensional chaotic maps

To extend the state space and increase non-linearity, a multidimensional PRNG (pseudo-random number generator) can be used. The well-known Lorenz-attractor system is an example of an unpredictable three-dimensional system:

$$f(x, y, z) = \begin{pmatrix} b & 0 & y \\ 0 & 0 & s \\ -y & -s & -1 \end{pmatrix} \begin{pmatrix} x \\ y \\ z \end{pmatrix} \tag{8.7}$$

The phase-space diagram (Figure 8.14) shows the Lorenz attractor, a fractal object, which has a bounded state space but a non-periodic trajectory (for a certain set of parameters). Multidimensional systems have still never been utilized in digital cryptography, owing to the complexity of the numerical integration and the non-uniform distribution of each variable.

8.4.3 Orbit lengths of chaotic finite-state systems

Floating-point implementations of chaotic maps can exhibit behaviour completely different from their continuous prototypes, e.g. very short cycles, depending on the particular numerical representation, the precision and the rounding or truncation procedure. All the trajectories of finite-state chaos generators are periodical orbits; see Figure 8.15. What are the minimal, typical and maximal periods of such orbits? Such questions are very important for long-stream PCNGs. In the best design one orbit interconnects all the states, and this orbit is unpredictable. When we consider a block cipher (which has a small number of first iterations), it is not so important to have the longest possible

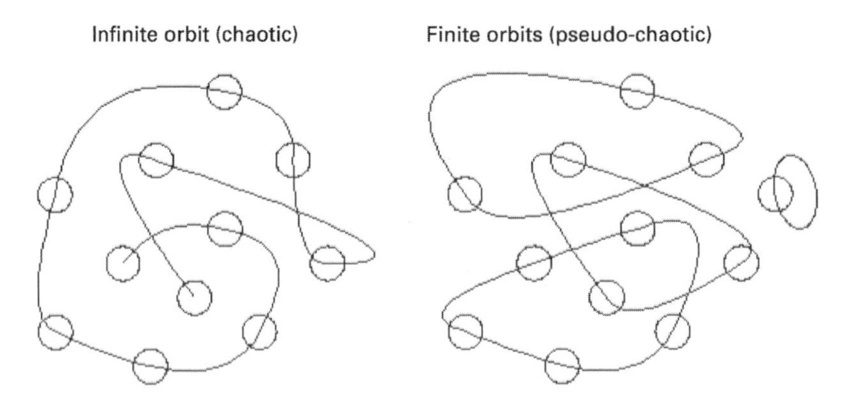

Figure 8.15 Orbits: on the left, a continuous chaotic orbit and on the right finite-state pseudo-chaotic orbits.

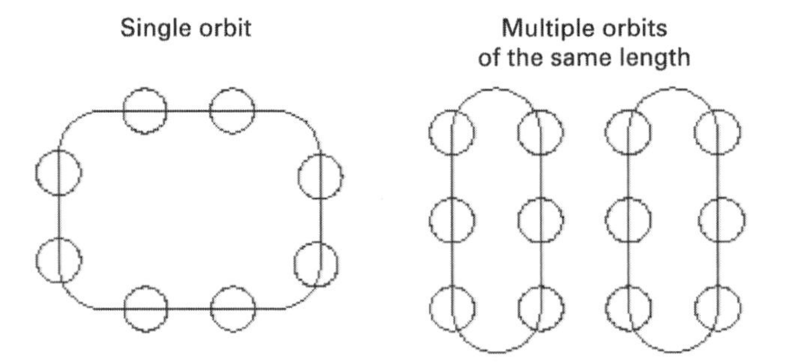

Figure 8.16 Preferable orbit configurations of PCNGs.

orbit; a block cipher can also operate efficiently on multiple orbits of the same length (see Figure 8.16).

The problem with floating-point PCNGs is that orbit lengths vary considerably. Cryptanalysts can predict short orbits and deliver a successful attack using the patterns obtained.

Very few chaotic continuous-state systems meet cryptographic requirements. Cambered chaotic maps, e.g. ($4rx(1-x)$, $\sin \pi x$) cause the output to have a particular non-uniform distribution. Only piecewise linear maps (the periodic tent map, the sawtooth map) are close to having pseudo-random properties. However, the FPA implementation of these maps produces short orbits and rounding-to-zero stops.

Discrete Kolmogorov flows do not replace the plaintext bits by random-like ciphertext but provide efficient permutations. In combination with substitution (confusion) transformations, Kolmogorov flows can be used in block ciphers.

Discrete chaotic systems such as the linear congruential generator, $x_{n+1} = Ax_n + C \bmod M$, the affine transformation $x_{n+1} = Ax_n + C$ and the generator of Blum *et al.* [40], $x_{n+1} = x_n^2 \bmod N$, are widely used in conventional cryptography.

8.4.4 Introducing the seed

There are three basic areas where the key k can be introduced:

- the initial-condition vector x_0
- the parameter vector c
- the shift n_0 (in other words, the start time t_0)

Generally the key k is a combination of two subsets and an integer:

$$k = \langle x', c', n'r \rangle, \qquad \text{where } x' \subseteq x_0, \quad c' \subseteq c, \quad n' = n_0$$

The basic idea in choosing $\langle x', c', n'r \rangle$ is to ensure a high complexity for predicting the key and the plaintext given either the final state x_{n_0+k} or the sequence $x_{n_0}, x_{n_0+1}, \ldots, x_{n_0+k}$; which of these is given depends on how the PCNG is used. We should define the space K (the space of allowed values of the key) in such way that for all $k \in K$ the PCNG generates statistically indistinguishable sequences.

Operation

The shift n_0 denotes the number of idle iterations before the PCNG starts producing a bit stream. The corresponding initial-condition vectors are

$$x_0, x_1, x_2, \ldots, x_{n_0}, x_{n_0+1}, \ldots$$

We have

$$z_i = g(x_{n_0+1}), \qquad i = 0, 1, 2, \ldots$$

where $g(x)$ is a partitioning function and z_i is the output bit or binary sequence.

The shift operation can be used to provide the hash property of the iterator. For sufficiently large n_0 and unknown parameter c, the inverse transformation

$$x_0 = v(x_{n_0})$$

such as $x_{n_0} = f^{n_0}(x_0)$, becomes computationally difficult.

An example of a PCNG

Let the sawtooth map

$$x_{n+1} = (px_n/q) \bmod$$

be the iterated function f. An example of a software PCNG given in pseudo-code follows.

```
class SawtoothKey {
    double p;
    double q;
```

```
    double seed;
    int shift;
} ... double iterateSawtoothMap(double x, SawtoothKey k) {
    return mod(k.p/k.q*x,k.q);
}

byte[] generateSeq(int length, Key k) {
    byte[] output[length];
    double x = k.seed;
    for(int i = 0; i < k.shift; i++) {   //idle cycles (shift)
        x = iterateSawtoothMap(x,k);
    }
    for(int i = 0; i < length; i++) {    //main cycles
        x = iterateSawtoothMap(x,k);
        output[i] = round(x);
    }
    return output;
}
```

8.5 Chaotic stream ciphers

Stream ciphers form an important class of encryption algorithms. As mentioned earlier, they encrypt individual characters (usually binary digits) of a plaintext message one at a time, using an encryption transformation that varies with time. By contrast, block ciphers tend to encrypt simultaneously groups of characters of a plaintext message using a fixed encryption transformation. Stream ciphers are generally faster than block ciphers in hardware and have less complex hardware circuitry. They are also more appropriate, and in some cases mandatory (e.g. in some telecommunications applications), when buffering is limited or when characters must be individually processed as they are received. Because they have limited or no error propagation, stream ciphers may also be advantageous in situations where transmission errors are highly probable.

8.5.1 Chaotic Vernam cipher

The Vernam cipher converts plaintext to ciphertext one bit at a time. A keystream generator outputs a stream of bits: $k_1, k_2, \ldots, k_n$. This keystream (sometimes called a running key) is passed through an XOR gate with a stream of plaintext bits, $p_1, p_2, \ldots, p_n$ to produce the stream c_i of cipher bits:

$$c_i = p_i \oplus k_i$$

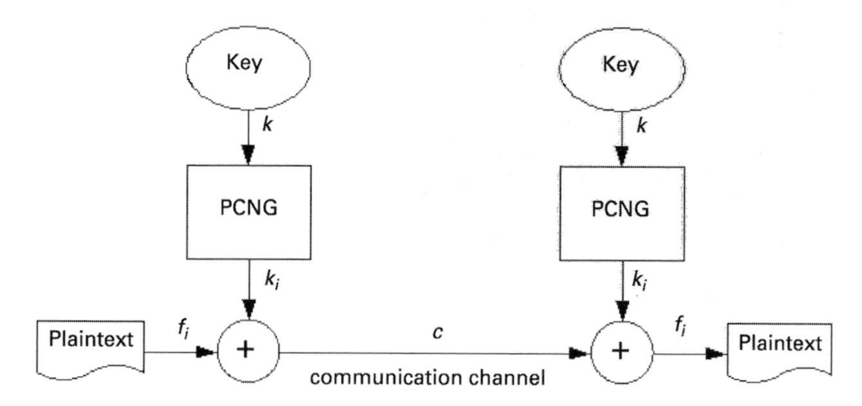

Figure 8.17 The chaos-based Vernam cipher.

At the decryption end, the ciphertext bits are XORed with an identical keystream to recover the plaintext bits:

$$p_i = c_i \oplus k_i$$

Since

$$p_i \oplus k_i \oplus k_i = p_i$$

this works nicely. For the chaos-based Vernam cipher, $k_1, k_2, \ldots, k_n$ is a chaotic sequence. Figure 8.17 shows the basic encryption scheme, which is based on a chaos generator. The secret key is a seed of the PCNG. Generally, the seed k is a triple of two vectors (the initial condition k_0 and the parameters c) and an integer (the shift n_0).

8.5.2 Security of a stream cipher

The system's security depends entirely upon the workings of the PCNG. If the PCNG emits a repeating pattern, the algorithm will have negligible security. In the case of the Vernam cipher, the length of the repeating pattern can be discovered by a procedure known as counting coincidences. As soon as the length is obtained, a pair of ciphertext bytes p_i corresponding to the same key k^* can be identified:

$$c_1 = p_1 \oplus k^*, \qquad c_2 = p_2 \oplus k^*$$

The sum of these ciphertext bytes equals the sum of the two plaintext bytes:

$$p_1 \oplus p_2 = c_1 \oplus c_2$$

Since ASCII text contains plenty of redundancy (for instance, English has 1.3 bits of real information per byte) the right combination of p_1 and p_2 can be easily estimated given $c_1 \oplus c_2$. But if the PCNG emits an endless non-repeating sequence of truly random bits, we have perfect security.

Shannon [38] proved that a necessary condition for a symmetric-key encryption scheme to be unconditionally secure is that $H(k) > H(p)$, where H is the entropy. That is, the uncertainty of the secret key must be at least as great as the uncertainty of the plaintext. If the key has bit length $l(k)$, and the key bits are chosen randomly and independently, then $H(k) = l(k)$ and Shannon's necessary condition for unconditional security becomes $l(k) = H(p)$. The one-time pad is unconditionally secure regardless of the statistical distribution of the plaintext and is optimal in the sense that its key is the smallest possible among all symmetric-key encryption schemes having the property $l(k) = H(p)$.

8.5.3 Multiple iterated functions

Initially proposed by Protopopescu [39], a composite PCNG consists of various different iterated functions (chaotic maps), which allows us to extend the key space and complexity. Each iterated function is seeded with a particular initial condition and parameters. A combination of the seeds forms a key. A function $\Sigma(x) \equiv \sum(x^1, x^2, \ldots, x^m)$ transforms the m streams into a single output. An example of $\sum$ is an XOR operation:

$$\sum(x) = \bigoplus_{1 \le i \le m} x^i$$

This approach complicates cryptanalysis *quantitatively*: it keeps the vulnerabilities of elementary generators. Each generator must ensure an almost uniform distribution, otherwise the combined output x will show a unique statistical pattern characterizing the set of iterated functions used.

8.5.4 Cryptanalysis based on Bayesian estimation

Maximum *a posteriori* (MAP) and maximum-likelihood (ML) estimations are fundamental techniques for recovering information from a noisy signal. Murphy *et al.* [37] showed that a number of successful cryptanalyses of block ciphers, including linear and differential analysis, can be regarded as being within the ML framework.

Bayes' rule

Bayes' rule (section 3.2) relates the conditional probability of 'B given A' to that of 'A given B':

$$\pi(B|A) = \frac{\pi(B)\pi(A|B)}{\pi(A)}$$

where A, B are events; $\pi(A)$, $\pi(B)$ are the total probabilities that events A and B respectively will occur; $\pi(A|B)$ is the conditional probability of A given B; $\pi(B|A)$ is the conditional probability of B given A.

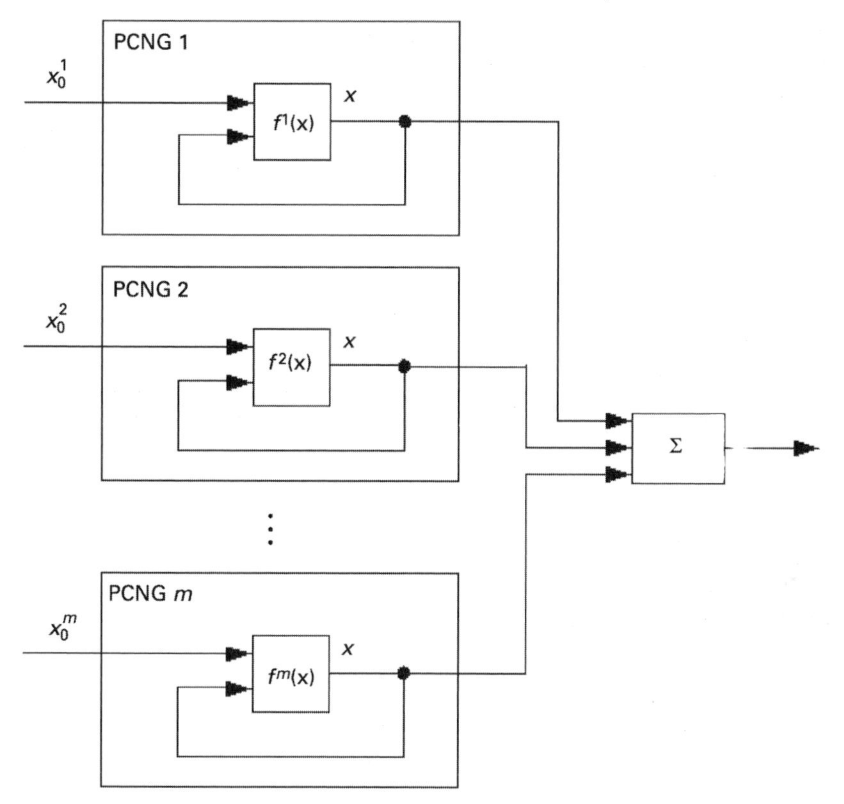

Figure 8.18 A composite PCNG.

In cryptanalysis Bayes' rule is written in the form

$$\pi(p|c) = \frac{\pi(p)\pi(c|p)}{\pi(c)}$$

where $p \in \{0, 1^n\}$ is the plaintext bit or bit block we want to recover from the cipher $c \in \{0, 1^n\}$.

8.5.5 Advantages and disadvantages of stream ciphers

We can summarize as follows. Basically, stream ciphers are considered as less secure than block ciphers because the complexity of the cryptanalysis entirely depends on the unpredictability of the pseudo-random generator, especially when the encryption function is a sample (XOR). By contrast, using block ciphers we can increase the complexity and non-linearity by means of multi-round substitution and mixing transformations.

The advantage of stream ciphers is that the cryptographic transformation depends on the state of the encryption system. That means each bit (byte or block) of the plaintext is encrypted into a different ciphertext. In a good stream cipher, $\pi(p|c) = \pi(p)$, i.e. there is no correlation between the ciphertext and the plaintext.

Multiply iterated functions do not provide a qualitative improvement in terms of cryptographic strength.

8.6 Chaotic block ciphers

Symmetric-key block ciphers are important elements in many cryptographic systems. Individually, they provide encryption or decryption of a data block. As a fundamental component, block ciphers allow the construction of one-way functions, PRNGs and stream ciphers. They may serve, furthermore, as a central component in message authentication techniques, data integrity mechanisms, entity authentication protocols and (symmetric-key) digital signature schemes.

8.6.1 Substitution and permutation

Substitution can be defined with a static S-box, i.e. a substitution table (in this case, the block size is limited) or a vector function $E : P \times K \to C$. A chaotic substitution transformation can be implemented in the following ways.

1. The chaotic map transforms a plaintext block (the plaintext is considered as the initial condition) into a ciphertext block. If the chaotic map is based on floating-point arithmetic then the algorithm should provide direct and inverse mappings from the plaintext bit sequence into floating-point numbers. If the chaotic map is based on binary arithmetic, the algorithm resembles traditional iterated block ciphers.
2. A PCNG generates an S-box, which is used in all rounds (iterations). The next block can be encrypted with the same S-box, or a new S-box can be generated (in this case the algorithm becomes a state cipher).

In both approaches, the key is a set of parameters of the chaotic system and/or a subset of the initial conditions.

Block encryption ciphers are designed based on a principle proposed by Shannon: non-linear mappings are extended by the use of *permutations* (mixing functions). Permutation does not affect the overall number of zeros and ones but, in combination with substitution, increases non-linearity and resistance to the various different cryptanalysis techniques. Figure 8.19 shows an iterated chaotic block cipher. Typically, permutations are achieved with a key-independent P-box (i.e. a permutation table) and/or bit rotations. In chaotic cryptography, a permutation can be implemented as a separate transformation, for example, using Kolmogorov flows, or together with substitution if the chaotic map ensures good mixing.

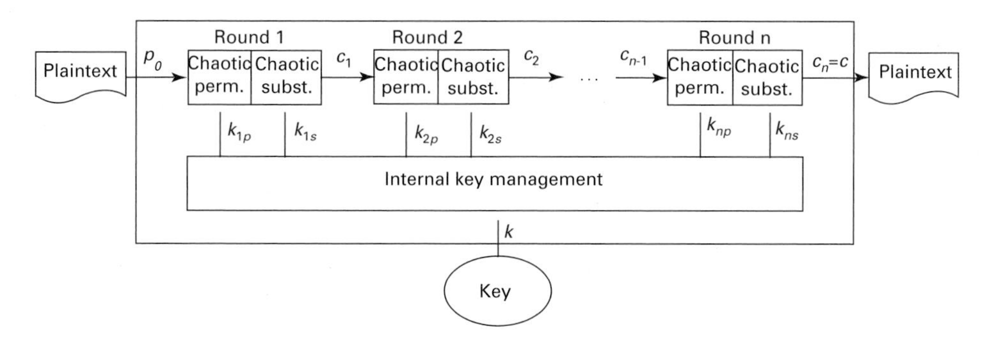

Figure 8.19 Iterated chaotic block cipher – substitution and permutation rounds (iterations).

8.6.2 Mode of operation

A block cipher encrypts plaintext in fixed-size n-bit blocks. For messages exceeding n bits, the simplest approach is to partition the message into n-bit blocks and encrypt each separately. This electronic codebook (ECB) mode has disadvantages in most applications, thus motivating other methods of employing block ciphers for larger messages.

8.6.3 Measure of diffusion

The word 'diffusion' has different meanings in different sciences. We have already described this term as it is commonly used in cryptology. Let plaintexts p, $p_1 \in P$, $p \neq p_1$, such that $\Delta p = p \oplus p_1$. In words, Δp is the 'difference' of the pair of plaintexts p and p_1. Let ciphers c, c_1 be such that $c = f(p)$ and $c_1 = f(p_1)$. In differential cryptanalysis, we try to answer the following questions.

- How far is $\Delta c = c \oplus c_1$ from Δp?
- If we fix the value of p_1 and change p from the set $\prod$, what are all the possible values of Δc?

Shannon stated: 'In a good mixing transformation . . . functions are complicated, involving all variables in a sensitive way. A small variation of any one [variable] changes [the outputs] considerably' [26].

What does a sensitive dependence on the initial conditions mean in chaos and in cryptology? In chaos it is assumed that the difference between two plaintexts Δ_p is small (infinitely small) and then the effect of the mapping (dynamics) is measured *asymptotically*, when time goes to infinity, using the Lyapunov exponents. In cryptography the effect of a single iteration of f is measured (or a number of iterations, if it is a round cipher) on *all possible values* of the difference Δ_p between two plaintexts; then the Hamming distance may be used as a metric to measure λ in discrete-state systems.

The most commonly used measure of diffusion in cryptology is called the *differential approximation probability* (*DP*) and it is defined as

$$DP = \max_{\Delta p \neq 0, c} P(\Delta c | \Delta p)$$

where

$$P(\Delta c | \Delta p) = \frac{\sharp\{p \in P | f(p) \oplus \Delta c\}}{2^m}$$

and m is the length of the encryption block in which the plaintexts are diffused. Kocarev [26] outlines the following properties of *DP*.

- $2^{-m+1} \leq P(\Delta c | \Delta p) \leq 1$.
- $P(\Delta c | \Delta p) = 1$ means that there exists a difference Δp that is always mapped by f to a difference Δc.
- If $P(\Delta c | \Delta p) = 2^{-m+1}$ then, for a given Δp, Δc is uniformly distributed.
- *DP* is a measure of differential uniformity of the map f: it is the maximum probability of having an output difference Δc when the input difference is Δp.
- Decreasing *DP* leads to an increase in the complexity of the differential attack.

8.6.4 Measure of non-linearity

The most commonly used measure of non-linearity in cryptology is called the *linear approximation probability* (*LP*). Kocarev [26] defined *LP* as follows:

$$LP = \max_{a, b \neq 0} [2P_{a,b} - 1)^2]$$

where

$$P_{a,b} = \frac{\sharp\{p \in P | p \bullet a = f(x) \bullet b\}}{2^m}$$

In the last equation, $a, b \in \{1, 2, \ldots, 2^m - 1\}$ and $\alpha \bullet \beta$ denotes the *parity of the bitwise product* of α and β. Although one always assumes that f is a non-linear mapping, it may happen that there exists a linear expression of f as an 'equation' for a certain modulo-2 sum of input bits of p and output bits of c:

$$p \bullet a \oplus f(p) \bullet b = 0$$

If the expression is satisfied with probability much more or much less than 0.5 then f can be approximated with a linear mapping. If, however, the expression is satisfied with probability close to 0.5 then f has 'strong' non-linear properties.

Kocarev [26] pointed out the following properties of *LP*.

- Let $N = \sharp\{p \in P | p \bullet a = f(x) \bullet b\}$. It easy to see that $0 \leq N \leq 2^m$.
- We write $A = a \bullet p$ and $B = b \bullet f(p)$.

  For $N = 2^m$, it follows that $A \oplus B = 0$ is satisfied for all p, which means that there exists a linear expression relating the input and output bits. In a similar way, $N = 0$ means that $A \oplus B = 1$ is satisfied for all p. In both cases $LP = 1$.

- For $N = 2^{m-1}$, the expression $A \oplus B = 0$ is satisfied with probability 1/2, and therefore $LP_{a,b} = 0$.
- LP is the square of the maximal imbalance of the following event: the parity of the input bits selected by the mask a is equal to the parity of the output bits selected by the mask b.
- Decreasing *LP* leads to an increase in the complexity of the linear attack.

8.6.5 Design approach

The general approach to block cipher design consists of the following 10 steps.

1. **Choose a state space** X (either binary sequences or floating-point numbers).
2. **Find a chaotic map** f mixing the plaintext and thereby providing an unpredictable transformation and producing a pseudo-random ciphertext.
3. **Introduce the plaintext, the key and the ciphertext into the chaotic system.** Define a valid subset of the input (plaintext, key) and output (ciphertext) in such a way that the system is structurally stable on these initial conditions.
4. **Choose a mode of operation.**
5. **Define the basic parameters of the block cipher:**
   (a) block size (l_b)
   (b) key size (l_k)
   (c) number of rounds (N_r)
6. **Evaluate the theoretical security for different parameter sets** (l_b, l_k, N_r) using the following criteria:
   (a) the pseudo-randomness of the output
   (b) the computational complexity of encryption and decryption (the class of computational problem, the number of operations required)
   (c) the entropy of the key space, $H(K)$. If all the combinations of l_k bits are valid and define a different bijection, $H(K) = \log_2 |K|$
7. **Evaluate the practical security for different parameter sets** (l_b, l_k, N_r), checking the resistance to known attacks that use
   (a) differential cryptanalysis (*DP*)
   (b) linear cryptanalysis (*LP*)
   (c) generalizations of differential and linear cryptanalysis
   (d) dedicated attacks applicable with a small number of rounds (i.e. cryptanalysis based on complete knowledge of the algorithm)
8. **Evaluate the performance and implementation complexity.** The algorithmic complexity affects the performance and the implementation costs both in terms of development and fixed resources.

9. **Evaluate the data expansion.** It is generally desirable, and often mandatory, that encryption does not increase the size of the plaintext data.
10. **Evaluate the error propagation**. The decryption of ciphertext containing bit errors may result in various effects on the recovered plaintext, including the propagation of errors to subsequent plaintext blocks. Different error characteristics are acceptable in various applications.

8.6.6 Conclusions

- The unpredictability of a cryptographic sequence, closely related to randomness and incompressibility, defines the theoretical complexity of cryptanalysis. Absolutely unpredictable sequences (having a truly uniform distribution) provide perfect security (one-time pad cipher).
- In practice, we use weaker analogues of unpredictability to identify a measuring instrument (a Turing machine or computer): *pseudo*-randomness, *computational* incompressibility, *algorithmic* complexity.
- Pseudo-randomness implies that the probability distribution of a given n-bit string can be distinguished efficiently (using polynomial-time computations) from the uniform distribution on $\{0, 1^n\}$. Unpredictability implies that there is no efficient probabilistic machine operating in polynomial-time that can predict the next bit (or subsequence) of the string. The two concepts are equivalent.
- The algorithmic complexity (or computational incompressibility) can be considered as a quantitative measure of unpredictability. Unfortunately, it is practically impossible to decide whether a given string is algorithmically random (i.e. to find the shortest algorithm generating the string).
- Pseudo-random number generators are based on a one-way function. The one-way function relies on a certain mathematical problem, to which an efficient solution has not been found. The algorithmic complexity of the existing (inefficient) solution is a measure of its unpredictability. 'Unsolved' problems applicable to cryptography can be found in discrete mathematics (integer factorization, RSA etc.) and in chaos theory (if the parameters are not known, it is difficult to predict the next state after one or more iterations or, vice versa, guess the previous states).
- The information entropy is another quantitative measure of unpredictability (pseudo-randomness). Practically, we estimate the pseudo-randomness by comparing the probability distribution of a cryptographic sequence with the uniform distribution. We should keep in mind that all finite-state generators produce periodical sequences. Consequently, a sufficiently long string, generated by a computer program, becomes predictable. To know the maximum length for an unpredictable subsequence is important when estimating the maximum length of plaintext string that can be securely encrypted with a single cryptographic sequence.

- Pseudo-random ensembles are unpredictable by probabilistic polynomial time machines associated with feasible computations but may be predictable by infinite-power machines. By contrast, chaotic systems are unpredictable by infinite-power machines (analogue computers) but may be predictable by probabilistic polynomial-time machines [26].
- Though truly chaotic systems can produce algorithmic random sequences, it is impossible to design such a generator on a computer. Only a certain 'level' of or approximation to randomness can be achieved in a chaos-based cipher; like all the conventional ciphers, chaos-based ciphers are far from giving the perfect security that is theoretically possible.

REFERENCES

[1] A. J. Menezes, P. C. van Oorschot and S. A. Vanstone. *Handbook of Applied Cryptology*. CRC Press, 1996. Available from WWW: `http://www.cacr.math.uwaterloo.ca/hac/`.

[2] B. Schneier. *Applied Cryptography*, 2nd edn. John Wiley and Sons, 1996.

[3] J. Gleich, *Chaos: Making a New Science*. Penguin, 1988.

[4] K. Clayton. Basic concepts in non-linear dynamics and chaos, 1996. Available from WWW: `http://www.vanderbilt.edu/AnS/psychology/cogsci/chaos/workshop/Workshop%F.html`.

[5] J. Mendelson and E. Blumenthal. Chaos theory and fractals [online], 2000. Available from WWW: `http://www.mathjmendl.org/chaos/`.

[6] G. Elert. Chaos hypertext book. 2001. Available from WWW: `http://www.hypertextbook.com/chaos/`.

[7] A. B. Cambel. *Applied Chaos Theory: A Paradigm for Complexity*. Academic Press, New York, 1993.

[8] B. B. Mandelbrot and J.W. van Ness. Fractional Brownian motions, fractional noises and applications. *SIAM Review*, **10**(4): 422–37, 1968.

[9] M. J. Feigenbaum. The universal metric properties of non-linear transformations. *J. Statistical Physics*, **21**: 669–706, 1979.

[10] R. Matthews. On the derivation of a chaotic encryption algorithm. *Cryptologia*, **13**: 29–42, 1989.

[11] T. Habutsu, Y. Nishio, I. Sasase and S. Mori. A secret key cryptosystem by iterating chaotic map. In *Proc. EUROCRYPT*, pp. 127–40, 1991.

[12] M. E. Bianco and D. Reed. An encryption system based on chaos theory. US Patent no. 5048086, 1991.

[13] M. S. Baptista. Cryptography with chaos. *Physics Letters A*, **240**(1, 2), 1998.

[14] W. K. Wong. Chaotic encryption technique, 1999. Available from WWW: `http://kitson.netfirms.com/chaos/`.

[15] H. Waelbroeck and F. Zertuche. Discrete chaos, 1998. Available from WWW: `http://papaya.nuclecu.unam.mx/~nncp/chaos98.ps`.

[16] D. D. Wheeler. Supercomputer investigations of a chaotic encryption algorithm. *Cryptologia*, **15**: 140–50, 1991.

[17] J. Fridrich. Secure image ciphering based on chaos. Technical Report, USAF, Rome Laboratory, New York, 1997.

[18] J. Scharinger. Secure and fast encryption using chaotic Kolmogorov flows. Technical Report, Johannes Kepler University, Department of System Theory, 1998. Available from WWW: http://www.cast.uni-linz.ac.at/Department/Publications/Pubs1998/Scharin%ger98f.htm.

[19] A. Fog. Chaotic random number generators, 1999. Available from WWW: http://www.agner.org/random/theory/.

[20] S. Hollasch. IEEE standard 754 floating point numbers, 1998. Available from WWW: http://research.microsoft.com/~hollasch/cgindex/coding/ieeefloat.html.

[21] M. K. Ho. Chaotic encryption techniques. Master's thesis, Department of Electronic Engineering, City University of Hong Kong, 2001. Available from WWW: http://www.ee.cityu.edu.hk/~50115849/fyp/project.htm.

[22] L. Cappelletti. An fpga implementation of a chaotic encryption algorithm. Bachelor's thesis, University Degli Studi di Padova, 2000. Available from WWW: http://www.lcappelletti.f2s.com/Didattica/thesis.pdf.

[23] D. Machleid. The use of chaotic functions in cryptography. Master's thesis, Computer Science WWU, Seattle, 2001. Available from WWW: http://www.jps.net/machleid/thesisp.html.

[24] H. Gutowitz. Cryptography with dynamical systems, 1995. Available from WWW: http://www.santafe.edu/~hag/crypto/crypto.html. ESPCI, Laboratoire d'Electronique, Paris.

[25] Z. Kotulski and J. Zczepanski. On the application of discrete chaotic dynamical systems to cryptography, dcc method. *Biuletyn Wat Rok XLVIII*, **566**(10): 111–23, 1999. Available from WWW: http://www.ippt.govol.pl/~zkotulsk/kreta.pdf.

[26] L. Kocarev. Chaos and cryptography, 2001. Available from WWW: http://rfic.ucsd.edu/chaos/ws2001/kocarev.pdf.

[27] C. M. Grinstead and J. L. Snell *Introduction to Probability*. American Mathematical Society, 2001. Available from WWW: http://www.dartmouth.edu/~chance/teaching_aids/books_articles/probability_book/book.html.

[28] L. Lovasz. Theoretical methods in computer science, 1998. Available from WWW: http://artemis.cs.yale.edu/classes/cs460/Spring98/contents.html.

[29] D. Lai, G. Chen and M. Hasler. Distribution of the Lyapunov exponent of the chaotic skew tent map. *Int. J. Bifurcation and Chaos*, **9**(10): 2059–67, 1999.

[30] N. Masuda and K. Aihara. Finite state chaotic encryption system, 2000. Available from WWW: http://www.aihara.co.jp/rdteam/fs-ces/.

[31] T. Erber, T. Rynne, W. Darsow and M. Frank. The simulation of random processes on digital computers: unavoidable order. *J. Computational Physics*, **49**: 349–419, 1983.

[32] T. Stojanovski and L. Kocarev. Chaos based random number generators, Part I: Analysis. *IEEE Trans. Circuits and Systems I*, **48**(3), 2001.

[33] T. Stojanovski and L. Kocarev. Chaos based random number generators Part II: Practical realization. *IEEE Trans. Circuits and Systems I: Fundamental Theory and Applications*, **48**(3), 2001.

[34] D. Knuth. *The Art of Computer Programming*, vol. 2 of *Seminumerical Algorithms*, 2nd edn. Addison-Wesley, 1981.

[35] I. Prigogine. *From Being to Becoming*. Freeman and Co., 1980.

[36] J. Fridrich. Symmetric ciphers based on two dimension chaotic map. *Int. J. Bifurcation and Chaos*, **8**(6): 1259–84, 1998.

[37] S. Murphy, F. Piper, M. Walker and P. Wild. Likelihood estimation for block cipher keys, 1995. Available from WWW: http://www.isg.rhul.ac.uk/~sean/.

[38] C. E. Shannon Communication theory of secrecy systems. *Bell System Technical Journal*, **28**(4): 656–715, 1949.

[39] V. Protopopescu. In *Proc. Topical Conf. on Chaos in Physical Systems*, Georgia Institute of Technology, 1986.

[40] L. Blum, M. Blum and M. Shub. A simple unpredictable pseudo random number generator. *Siam J. Computing*, **15**, 364–83, 1986.

Index